Fluid Power
with
Microprocessor Control:
An Introduction

Fluid Power with Microprocessor Control: An Introduction

EDWARD W. REED

Senior Lecturer, Department of Mechanical and Production Engineering, Leeds Polytechnic

IAN S. LARMAN

Principal Consulting Engineer, Island Associates

Prentice / Hall   PHI   International

Englewood Cliffs, New Jersey   London   Mexico   New Delhi
Rio de Janeiro   Singapore   Sydney   Tokyo   Toronto   Wellington

Library of Congress Cataloging in Publication Data

Reed, Edward W., 1938–
  Fluid power with microprocessor control.

  Bibliography: P
  Includes index.
  1. Fluid power technology.   2. Microprocessors.
  I. Larman, Ian S., 1937–   . II. Title.
  TJ843.R4   1985   621.2   85–12050
  ISBN 0–13–322488–0
  ISBN 0–13 322470–8 Pbk

British Library Cataloguing in Publication Data

Reed, Edward W.
  Fluid power with microprocessor control.
  1. Fluid power technology——Data processing
  2. Microcomputer
  I. Title   II. Larman, Ian S.
  620.1′06′0285404   TJ840

  ISBN 0–13–322488–0
  ISBN 0–13–322470–8 Pbk

© **1985 by Prentice-Hall International, UK, Ltd**

Prentice-Hall, Inc., *Englewood Cliffs, New Jersey*
Prentice-Hall International, UK, Ltd, *London*
Prentice-Hall of Australia Pty Ltd, *Sydney*
Prentice-Hall Canada, Inc., *Toronto*
Prentice-Hall Hispanoamericana, S.A., *Mexico*
Prentice-Hall of India Private Ltd, *New Delhi*
Prentice-Hall of Japan, Inc., *Tokyo*
Prentice-Hall of Southeast Asia Pte Ltd, *Singapore*
Editora Prentice-Hall do Brasil Ltda, *Rio de Janeiro*
Whitehall Books Ltd, *Wellington, New Zealand*

ISBN 0-13-322488-0
ISBN 0-13-322470-8 PBK

Photoset by Gecko Limited, Bicester, Oxon
Printed in Great Britain by
A. J. Wheaton & Co. Ltd, Exeter
1 2 3 4 5 6 7 8 9

Contents

Appendixes

Preface

Fluid power systems have undergone a dramatic change in recent years in that the majority of applications now demand the control of speed and force, as well as position, using electronics. With the development of the microcomputer, a whole new area of control is possible. It is generally felt that the combination of microelectronics to supply the 'brain' and fluid power to supply the 'brawn' is the most effective way of exploiting the benefits of fluid power systems.

The enormous potential for controlling mechanical elements in such sophisticated applications as flexible manufacturing and robotics, as well as simple everyday applications such as a tractor pulling a plough, gives some indication of the range of applications for the combination of microelectronics with fluid power.

This book was written with the practicing engineer in mind, as well as the undergraduate. It deals with the latest developments in electrically modulated hydraulic equipment, as well as introducing the complexity of microprocessor controls in a simple way. The authors work as consultants in the field of applied microelectronics and are independently registered with the Department of Industry in this capacity. They have used the information in this book on numerous occasions when conducting seminars and training courses for companies who are either actively engaged in manufacturing fluid power components or who are applying these components.

As this work is an introduction to the subject, the authors have not dealt in depth with any of the disciplines, viz. electronics, microcomputing, and fluid power, but there is sufficient information for real applications to be accomplished.

The elements of a complete fluid power system can be divided into the following subsystems: energy supply, control, motion and power, and feedback, and these elements are now considered.

Energy supply produces a fluid under pressure which is used to provide movement and force. This can be achieved by either a hydraulic pump or an air compressor.

Control of fluid power must be effective and safe. Microcomputers are undoubtedly the most suitable means for controlling fluid power systems of any complexity. They are inexpensive and powerful, and are able to act on information they receive in a far more sophisticated way than can previous methods of control.

Motion and *power* are obtained by using cylinders and motors. To give direction, speed and load control, valves are needed to regulate the flow of fluid in any given direction.

Feedback is the signal, from the components which monitor the system behavior, which verifies the execution of the input commands. Electronic feedback components used with microcomputers produce a degree of resolution hitherto unknown.

The microprocessor chosen for the basis of the illustrations is the Zilog 80 (Z80) as this is the microprocessor most widely used for control applications.

The first chapters of this book are a review of the various types of fluid power components which are currently available and form the basis of an introduction to the undergraduate in this area of control engineering. The predominance of hydraulics at the expense of pneumatics is deliberate because of the fine degree of control which can be achieved with hydraulic systems.

The manufacturers of pneumatic equipment led their counterparts in the hydraulic industry with the introduction of microelectronics. This was because pneumatics engineers have always had direct involvement with control circuits and appreciated the needs of manufacturing industry.

Chapter 4, as an introduction to the operation of microcomputers, shows in simple terms how the various components of a microcomputer system operate. It is intended to provide a basis for the many good specialist books which the microprocessor manufacturers supply.

The following two chapters are devoted to feedback and interfacing. They describe the 'nuts and bolts' as it were of computer controls. In the past, with analog systems for fluid power control, a detailed knowledge of electronic systems was needed. Using microcomputers, control becomes little more than a matter of verification of the movement of actuators.

Electrical noise is a source of problems for many engineers engaged in control. Chapter 7 shows the different ways of preventing this noise from corrupting microcomputer control systems.

An introduction to programming is given in chapter 8, comparing the merits of machine code programming with a higher level language. It is assumed that the reader has some knowledge of the programming language BASIC as it is widely used in home computers and taught in many schools. This chapter introduces the first of the worked examples for the control of a simple fluid power system.

Chapters 9 and 10 deal with hydraulic valves which are electrically modulated and are generally used to control mechanical devices to a high degree of accuracy. The current state of the art relating to these types of component is given, with information relating to future developments.

The low cost proportional control valve is now being employed in applications where previously servo technology was incorporated. Chapters 11 and 12 deal with worked examples. The case study in chapter 12 describes the operation of a hydraulic press which involves the control of three simultaneous actions. This example embodies the principles of control and the fluid power components which could be used in other applications such as flexible manufacturing and robotics.

The authors should like to acknowledge the help and assistance given by members of staff at Leeds Polytechnic in compiling this book.

Acknowledgement and thanks are made to the following companies for their invaluable help and permission to use their material:

Robert Bosch GmBH: pages 120, 122 and Appendix A.

G. L. Rexroth Ltd: pages 49, 128, 129.

Fenner Fluid Power (Division of J.H. Fenner & Co. Ltd): page 44.

Danfoss Ltd: pages 9, 50.

K.L.B. Controls (Holland): page 117.

The consideration, encouragement and substantial help from the author's wives, Diana Reed and Dorothy Larman, should not go without acknowledgement and thanks.

E.W. Reed
I.S. Larman

1

Pumps and Compressors

The function of hydraulic pumps and air compressors is to provide a fluid under pressure which will perform work. This chapter gives a brief description of pumps and air compressors and their principles of operation. It does not give an extensive description as there are various books available which cover this subject, but it is sufficient to provide an understanding of the principles of operation for working purposes.

1.1 OPERATION OF PUMPS AND COMPRESSORS

Hydraulic pumps and air compressors take in fluid at the inlet port at atmospheric pressure and give out fluid under pressure at the oulet port. The technique used is to expand the volume of space inside the pump or compressor and to fill the expanded volume with a fluid.

The principle of operation of hydraulic pumps is that atmospheric pressure causes a fluid to be drawn through the inlet port, sometimes called the 'suction port', into the pump. A subsequent reduction of the volume inside the pump causes the pressure to rise as it flows from the outlet.

Unless the flow of oil from the pump is released either by moving a load or by passing through a relief valve to the reservoir, serious damage can occur in the pump and relative pipework.

The technique of expanding and compressing volumes occurs in the operation of the air compressor. Atmospheric air is drawn into the compressor as a result of a pressure drop developing inside the compressor. Reducing the volume causes an increase of pressure, within the compressor, and this air, at an increased pressure, can be used to move a load.

There is a great difference between the behavior of oil and air with respect to compressibility. As an example: to achieve a pressure of 7 bar, atmospheric air must be reduced in volume by 7/8ths, whereas oil has only to be reduced in volume by 1/2000th to produce the same increase of pressure.

Only in systems in which very large volumes of oil are stored under pressure does the compressibility of oil need to be taken into consideration.

1.2 HYDRAULIC PUMPS

Hydraulic pumps are described below and come in various sizes. They produce flow rates between $1\,cm^3$ per revolution to $2\,l$ per revolution.

Different types of industry choose the design of pump which is most suited for the particular working environment. For example:

1. The machine tool designer opts for the *vane* pump because of its low noise characteristic of, say, 70 dBA (the noise level dBA is the decibels of amplitude which is weighted to correspond to the response of the human ear). The ability of vane pumps to deliver a variable flow of oil at a constant pressure is also of benefit to the machine tool designer. The action of providing a variable flow prevents heat build-up as would occur from, say, a fixed displacement pump like a gear pump.

2. The construction engineer favors the *piston* pump because of its ability to transmit greater power for its size than any other type of pump. Gear pumps are also used in the construction industry, sometimes as a part of the piston pump or as an alternative to the piston pump when large amounts of power are not used.

3. The agricultural engineer prefers the *gear* pump because of its low cost and robustness.

This list is used as an example to show individual preferences and does not presuppose that other pumps are not used according to the personal preference of the individual designer or the dictates of the different applications.

Each type of pump (vane, piston and gear) has its own design and application features. To appreciate how each pump can be used, it is necessary to consider the design features.

1.2.1 Vane Pumps

Vane pumps are used mainly for their low noise characteristics and will move large or small volumes of oil. Their main limitation is that dirt quickly reduces their efficiency by damaging the pressure-sealing of the vane tips. In fact, the sealing action of a vane pump is so searching that it is used by the major oil manufacturers to test the lubricating properties of their hydraulic fluids.

It is possible to produce vane pumps which have a fixed (Fig. 1.1) or variable flow control.

The vanes are pushed into the rotor by the cam ring as the rotation of the pump occurs. This reduction in volume creates an increase in pressure inside the pump body. This increase in pressure occurs until the vanes are deflected the maximum distance into the rotor. Further rotary movement allows them to be released from their grooves, either through the action of a spring or by hydraulic pressure. The subsequent increase in volume creates a pressure drop inside the pump. Oil is then forced into the pump by atmospheric pressure acting over the area of oil in the reservoir.

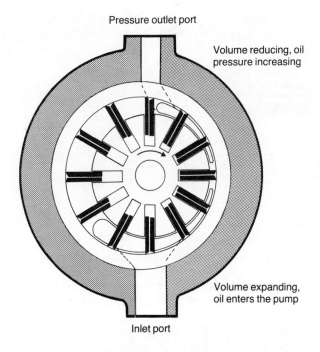

Fig. 1.1 Fixed vane pump.

The vanes are held on to the outer surface of the glide ring by either springs or hydrostatic forces.

1.2.2 Piston Pumps

Piston pumps are designed to produce high pressures of up to 700 bar (10 000 psi) and will in certain instances be very dirt-tolerant. The total swept volume of the pump is called its displacement. Piston pumps operate in two principal ways.

1. AXIAL PISTON: axial piston pumps have their pistons attached either to the rotating shaft or on the static control plate, called the slipper plate. In the latter case the pistons glide axially on slipper pads. The variable pump can deliver different flows by adjusting the angle at which the pistons run in relationship to the drive shaft, or the angle at which the fixed plate relates to the center of the drive shaft. The principle is shown in Fig. 1.2.

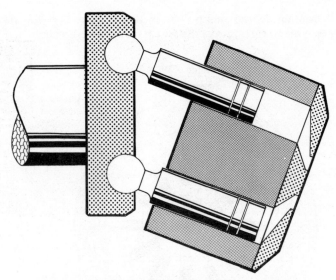

Fig. 1.2   Bent axis piston pump showing just two pistons.

2.   RADIAL PISTON: radial piston pumps (Fig. 1.3) provide either a fixed or variable flow ouput. The pistons are either fixed to the body of the pump and move by a camshaft attached to the drive shaft, or are driven by the drive shaft and move in an eccentric housing similar to that of a vane pump.

As the drive shaft rotates, the piston moves further into the pump body. The volume decreases, producing an increase of pressure. The movement of the piston into the body displaces an amount of oil proportional to the volume derived from the piston area and the displacement. This displacement is transferred from the piston housing into the hyraulic system.

Noise is one of the main objections to the use of axial piston pumps. On some of the larger sizes of pump, noise levels typically of 85–90 dBA are developed. However, radial piston pumps, which usually have pistons of larger sectional area and short strokes, tend to be much quieter. The maximum swept volume of a radial piston pump, however, will be, say, $250\,cm^3$ $(15\,in^3)$ per revolution, whereas the volume of an axial piston pump can reach up to $2\,l\,(122\,in^3)$ per revolution.

1.2.3   Gear Pumps

Gear pumps are all fixed displacement pumps. They are designed to operate with either an internal or an external gear drive. To provide increasing and decreasing volumes the gears must be enclosed inside a body or housing. There are two types of gear pump:

1.   internal gear pumps;
2.   external gear pumps.

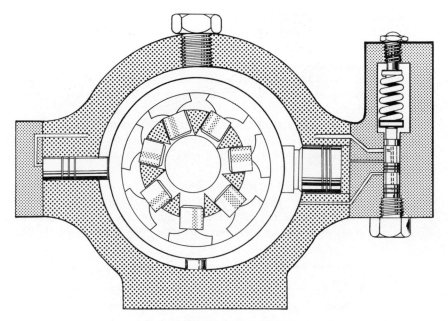

Fig. 1.3   Radial piston pump.

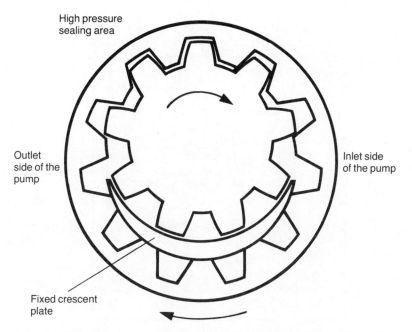

Fig. 1.4   Internal gear pump. The fixed cresent plate provides a sealing action between the inlet and outlet ports.

Internal gear pumps (Fig. 1.4) tend to be quieter running and are designed to compete with vane pumps. Noise levels of, say, 60dBA are possible with internal gear pumps but they are more expensive.

The pump comprises an inner gear ring and an external gear pinion. The gear pinion is held in mesh with a crescent-shaped wear strip. The action of the pressure loading on the intermeshing gears forces the pinion on to the wear strip and creates a seal.

Rotating the pinion in a clockwise motion expands the clearance between the teeth at the right-hand side of center and creates a pressure drop. The meshing of the teeth to left of center reduces the clearance and creates a pressure rise.

The drive shaft of the pump sometimes incorporates the gear pinion and is held in its correct position by the bearings mounted into the end covers. The internal gear ring rotates inside the body of the pump. One difficulty is to maintain the seal between the faces of the meshing gears. It is

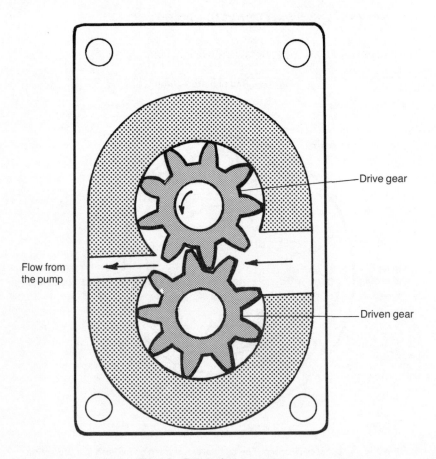

Fig. 1.5   External gear pump.

normal to have pressure plates which are held in position by the increasing pressure created by the pump.

External gear pumps (Fig. 1.5) are constructed from two gear pinions meshing together within a housing. As the teeth close together the volume reduces, creating a pressure rise. The opening action of the teeth creates an increase of volume and a pressure drop; thus oil is forced into the pump.

Gear pumps are compact in size and capable of transmitting a large amount of power. They are also of fixed displacement and therefore need to be sized for the desired flow requirement of the application. The speed of the electric motor or internal combustion engine is multiplied by the pump displacement to give the desired output flow.

External gear pumps normally produce about a 75 dBA noise level when running at maximum pressure and maximum speed. They are low cost and are suitable for pressures up to 300 bar (4350 psi).

1.3 VARIABLE FLOW HYDRAULIC PUMPS

Variable flow pumps alter the output of oil without changing the running speed. As previously discussed, flow is the product of the pump displacement multiplied by the driven speed. Therefore, to vary the pump flow it is necessary to alter the pump's displacement.

1.3.1 Discussion

Variable flow pumps have been used for a number of years to provide a compensation of pressure within a hydraulic system. They are produced in two design types, namely:

1. vane construction;
2. piston construction.

Each of these construction types can be produced to vary the flow rate and at the same time maintain the pressure setting of the system. In this configuration they are called 'pressure-compensating pumps'.

The system pressure acts on the control piston which overcomes the spring offset force and changes the stroke of the pump which then reduces the output flow of oil. The spring action can be from a mechanical or hydraulic source.

1.3.2 Variable Pumps: Vane

Variable vane pumps (Fig. 1.6) can be produced as variable flow pumps which compensate the flow to a set pressure. The cam or stroke ring is moved from the offset or eccentric position to run concentrically. They are available with either a spring or a hydraulic piston to offset the stroke ring and produce the pumping action.

For hydraulic control it is normal for the offset to be produced by a large piston and a smaller piston used as the 'spring'.

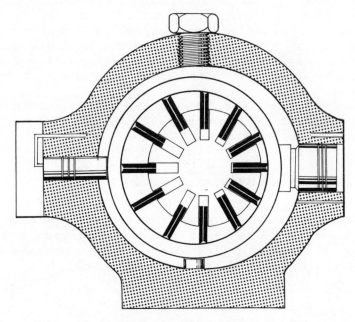

Fig. 1.6   Variable vane pump.

Hydraulically operated variable vane pumps operate in the following way. Oil from the high pressure port is fed internally to both the large and the small pistons. The pressure to the large piston operates through a valve which reduces the pressure to the desired level or setting. When the set pressure is reached, the pressure in the large piston escapes into the pump casing and the force of the small piston moves the cam or stroke ring to a position where the vanes run concentrically inside the cam ring. Provision is made to return the oil which leaks into the pump casing back to the reservoir.

1.3.3   Variable Pumps: Radial Piston

The operation described for vane pumps is exactly the same as it is used for radial piston pumps (Fig. 1.7).

The input of oil from the reservoir is directed to the pistons through the pintel shaft of the pump. Shoes glide on the cam ring and are attached to the pistons. Pressure oil is fed through the pistons to a pocket to lubricate the cam ring.

The piston housing is rotated by the drive shaft and produces the pumping action against the cam ring. The throw of the cam ring out of center determines the stroke of the piston.

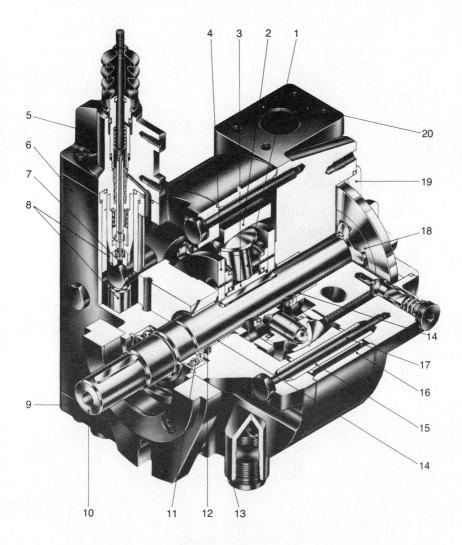

Fig. 1.7   Variable pump radial piston. 1 pistons with slippers, 2 control ring, 3 cylinder barrel, 4 balance plates, 5 control (manual servo), 6 front flange with control housing, 7 control pin, 8 control pistons, 9 control cover, 10 drive shaft, 11 ball bearing (absorbs radial loads), 12 shaft seal, 13 drain connection, 14 slide bearing (absorbs radial forces), 15 pivot pin, 16 spacer ring, 17 assembly bolts, 18 charge pump (gear pump), 19 end cover, 20 rear flange with connection ports.

1.3.4   Variable Pumps: Axial Piston

Axial pumps come in two different types, swash plate and bent axis designs, the difference being that the pistons are in a housing which rotates either on the center-line of the drive shaft or bent as approximately 25° to the shaft.

It is also possible to provide intermediate control of the oil flow by using a cylinder to control the offset of the pump. This controller moves the swash plate or bent axis of the pump to provide the variable flow output.

The swash plate pump is the simplest to manufacture in that the body housing of the pistons is in constant contact with the body of the pump, and therefore the inlet and outlet oils easily pass to and from the pistons. With bent axis pumps the inlet and outlet oils have to pass through the trunnions which are the means by which the pump moves to provide the variable flow output.

1.3.5   Control of Variable Pumps

There are numerous types of control action for variable pumps. The main types are:

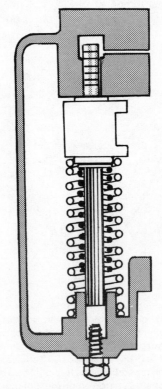

Fig. 1.8   Constant power controller.

1. constant power control;
2. constant pressure control;
3. constant flow control;
4. servo control;
5. electrical control;
6. manual control.

Having available such a wide choice of pump controllers for variable flow pumps means that the hydraulic system designer can use the best type of controller for a specific application. For example, a constant pressure control is used frequently for wheel-drive systems in off-highway vehicles.

1.3.5.1 Constant Power Control

The constant power controller (Fig. 1.8) uses two or three springs to offset the delivery angle of the pump. The springs can approximately provide the characteristic of the power requirement.

The power needed by a pump is dependent upon the speed and force requirements of the hydraulic actuator. These are supplied by the pump in the form of pressure and flow. The power curve is parabolic, since it follows the square law. The pump designer has to produce a similar characteristic with springs. The approximation of the power curve that the springs produce is shown by Fig. 1.9. The ideal curve to suit the maximum power draw-off

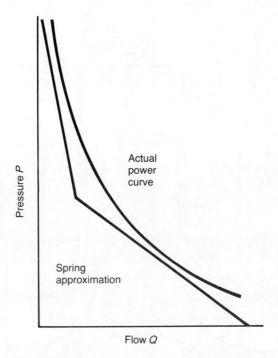

Fig. 1.9 Spring approximation curves compared with the actual power requirement.

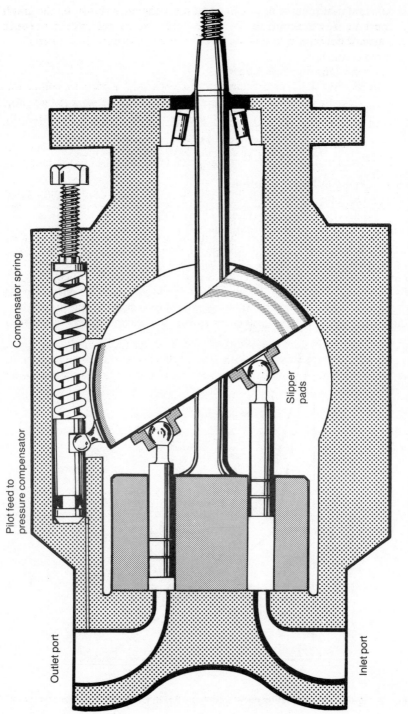

Compensator spring

Pilot feed to
pressure compensator

Outlet port

Slipper
pads

Inlet port

Fig. 1.10 The constant pressure controller fitted to a swash plate pump.

from the electric motor or internal combustion engine is shown by the graph on the right. The reaction to the force produced by the system pressure acting against the pilot piston is shown as having two distinct linear parts.

1.3.5.2 Constant Pressure Control

The method used for constant pressure control, or pressure compensation as it is generally called, is to offset the swash plate with a spring (Fig. 1.10). Opposing the spring is a hydraulic piston or cylinder to which the system pressure is fed via an orifice or restriction.

When the pressure within the hydraulic system produces sufficient thrust from the piston to compress the spring then the offset of the pumping action is reduced to zero.

When the pressure within the hydraulic system decays, the force of the spring overcomes that of the piston and provides once more an offset for the pump.

When the swash plate is at 90° to the center-line of the drive shaft there is no pump flow. Leakage of oil, through the manufacturing clearances produce a drop in pressure, and the constant pressure controller, or 'compensator' as it is sometimes called, then moves to generate flow to make up this loss.

1.3.5.3. Constant Flow Control

This control system uses a throttle device to set the desired flow output from the pump. Pressures from both sides of this throttle are then compared. This controller thus maintains a constant pressure drop across the throttling device (Fig. 1.11).

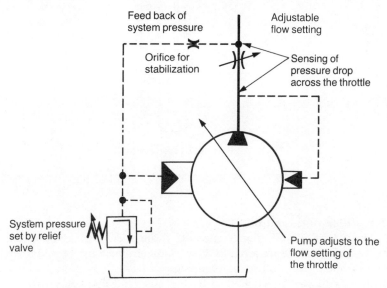

Fig. 1.11   Constant flow compensation shown diagrammatically.

A constant pressure controlled pump with a lighter offset spring is used. The load pressure from the outlet side of the throttle is returned to the spring offset side of the pump through a small orifice of approximately 0.8 mm diameter.

The pilot relief valve sets the system pressure. The spring in the compensator valve maintains a constant pressure drop across the throttle device, which in turn provides a constant flow regardless of any load variation. If a sharp-edged throttling device is used, the flow remains constant regardless of temperature fluctuations affecting the viscosity of the oil.

1.3.5.4 Servo Control

The swash plate, or piston housing in the case of bent axis pumps, is fixed to the cylinder and will move freely without any spring bias. A servo valve (see chapter 9) directly attached to the cylinder will cause the output of the pump to respond in direct relationship to the input value of the electrical signal (Fig. 1.12).

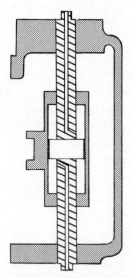

Fig. 1.12   Hydraulic servo control.

This type of control is used when rapid reversal of flow is needed for precise control.

1.3.5.5. Electrical Control

The electrical controller is operated through an electric motor and a screw thread. The desired pump flow is determined by the number of rotations made by the electric motor and the travel made by the nut. The nut is attached to the swash plate of the variable pump and alters the stroke of the pistons (Fig. 1.13).

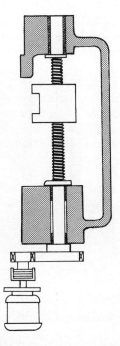

Fig. 1.13   Variable flow control using an electric motor drive.

It is normal with this type of controller to have a clutch fitted between the motor and the reduction gear-drive to the screw thread. This protects the motor should the nut come into contact with the end stops. Switches are usually fitted to prevent this happening.

1.3.5.6.   Manual Control

The manual controller acts in a similar way to the electrical controller, but instead of the electric motor moving the nut along the thread a hand wheel is provided (Fig. 1.14). It is possible to remove the hand wheel and provide a coupling to the operating screw. This can then be turned either remotely or through a drive mechanism.

1.4   AIR COMPRESSORS

Air compressors convert mechanical energy into the potential energy stored in compressed air. Compressors are manufactured in three types:

1.   vane;
2.   piston;
3.   screw or gear.

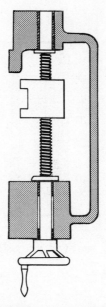

Fig. 1.14   Manual control of the pump flow.

The principle of operation of compressors is similar to that of hydraulic pumps. An initial large volume of air is squeezed into a smaller space, raising the pressure. The compressed volume of air is stored at the increased pressure in a receiver from which a supply can be drawn at a constant rate.

As a guide, air compressors require 6.5 kW of power to produce 1 m³ of air at 7 bar (or 1 horsepower to produce 4 ft³ of air at 100 psi).

1.4.1   Vane Compressors

Vane compressors (Fig. 1.15) have a series of spring loaded vanes which rotate either eccentrically, in a concentric housing, or concentrically, in an eccentric housing.

They are similar in construction to vane pumps but larger in size. They are used for installations where they are sited near the operator of the equipment or where low noise levels must be achieved.

1.4.2   Piston Compressors

Piston compressors are constructed similarly to internal combustion engines. A crankshaft to which connecting rods are attached is rotated. Pistons are fitted to the connecting rods and move up and down within cylinders. Valves are fitted to allow air into and out of the cylinder. Figure 1.16 shows this function.

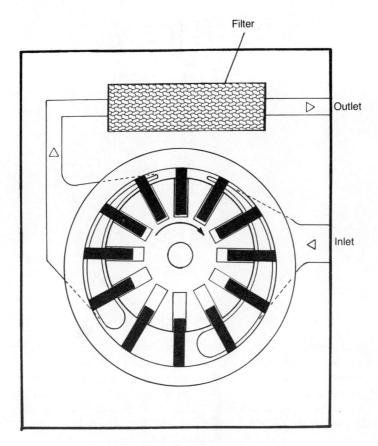

Fig. 1.15 Vane air compressor.

As the volume expands, the pressure drops and air at atmospheric pressure is drawn into the compression chamber through an inlet valve. When the volume of the cylinder decreases, by the action of the piston moving upwards, the pressure rises (provided both the valves are closed).

The increased pressure is directed through another valve to the cooler and storage reservoir.

Most people have seen piston compressors because they are frequently used by the construction industry when working on roads. It is possible, however, to find other types of compressor used in the construction industry.

1.4.3 Screw Compressors

Screw compressors (Fig. 1.17) consist of two or three helical gear hubs meshing together in a housing. Air from the surrounding atmosphere is drawn in and compressed against the compressor body.

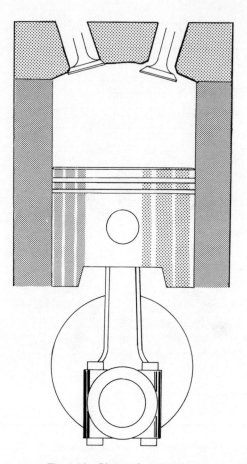

Fig. 1.16   Piston air compressor.

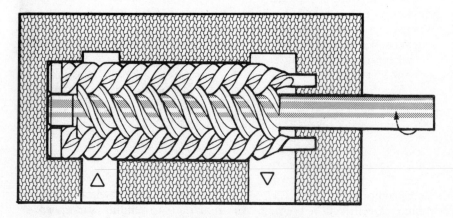

Fig. 1.17   Rotary screw compressor.

This type of compressor is perhaps the most popular for factory-based applications. It is quiet-acting, like a vane compressor, and is now becoming widely used for pneumatic applications.

1.5 WATER VAPOR PROBLEM

Free air, which is made up from the atmosphere, can be described as a mixture of gases. One of the gases is steam or water vapor.

All types of air compressor generate heat as they pump volume upon volume of air from their surrounding environment. In the process of compressing the air, the moisture content is also increased per unit volume.

Condensation takes place at the cooling stage and this moisture must be removed before storage in the reservoir, otherwise corrosion takes place. The way to achieve this is to pass the air, after it has been compressed, through a cooler or chiller. As the temperature drops, the saturated air releases droplets of excess moisture until the air is saturated at the lower temperature. Raising the temperature of the air again to that of the lowest ambient temperature means that the air can be stored in a dry condition. This is because the saturation point of air is greater at higher temperatures than at lower temperatures.

To ensure consistent control of a pneumatic system the compressed air must be treated to remove as much moisture as possible.

QUESTIONS

1. Explain the different basic types of hydraulic pump and state a charactersitic of each one.

2. Hydraulic pumps can compensate for pressure loads; what is pressure compensation and how is it achieved?

3. Give a method of producing constant flow control for a hydraulic pump.

4. What pump would be used for the following system requirements:
    (a) fast reversal of the flow;
    (b) where damage could occur to an internal combustion engine if too much power is being used;
    (c) where precise flow is required from a remote location?

5. Explain the different methods of compressing air.

6. What effect does a humid day have on compressed air?

FPMC-C

2

Valves

Valves are necessary to control the flow of fluid under pressure, to prevent damage to a fluid power system and to give direction of movement to an actuator. This chapter describes the different types of valve available for hydraulic and pneumatic systems. The valves are classified in four different categories. Check valves could have been considered to be in the category of directional control valves, but because some of their features are different from these valves they have been described separately.

2.1 INTRODUCTION

In simplified terms, a valve is a mechanical device which allows a fluid to flow through it when certain conditions are satisfied.

There are four main categories of valve design:

1. check valve;
2. throttle valve;
3. pressure control valve;
4. directional control valve.

All of these valves have their own particular features and are used as components in circuits to give a specific control function.

1. The *check valve* is designed to allow oil to flow in one direction only. It is used in a system to maintain the load for long periods of time and also to retain pressure in various parts of a circuit.
2. The *throttle valve* is used to control the speed of the actuator so that the desired velocities of the load or the fluid within the piping can be maintained. This type of valve comes in two designs: (a) simple restriction and (b) pressure and temperature compensated type.
3. The *pressure control valve* is always present in some form in a fluid power circuit. It limits the force exerted on the load by the actuator. It acts as a safety feature to limit the internal forces within the system as a whole.
4. The *directional control valve* provides a simple switching of the fluid to produce forward and backward movements of the actuator.

2.2 CHECK VALVES

Check valves are usually described as directional control valves but their features are so special that this chapter deals with them as a separate product type. They are very rarely used on their own as a directional control valve.

A check valve (Fig. 2.1) is essentially a ball or poppet which is held on a metal or plastic seat by a spring. It allows a force, created by the pressure of the fluid acting over the area of the seat, to move the ball against the spring. Flow then occurs from inlet to outlet. This valve allows flow in a single direction. Pressure at the spring side of the ball holds the valve closed with a force which is greater than that created by the system pressure.

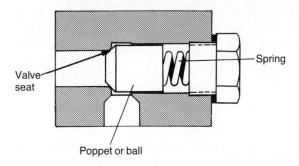

Fig. 2.1 Simple check valve.

The check valve is used when the pressure has to be maintained within the system. There is no way of automatically relieving this pressure without incorporating another valve.

2.3 PILOT-OPERATED CHECK VALVES

Pilot-operated check valves (Fig. 2.2) are similar to check valves but have a hydraulic or pneumatic pilot piston which unseats the ball or poppet. Fluid that is locked on the spring side is allowed to flow in the reverse direction by the action of this pilot piston.

This valve is used to hold a load for extended periods. To release the load, external fluid pressure must operate the pilot piston.

On double-acting cylinders the pilot pressure is usually taken from the unloaded side of the cylinder. When pressure is applied to the unloaded side of the cylinder to retract it, pressure builds up in the pilot piston chamber until the valve opens. Sometimes the load causes the cylinder to return faster than the desired speed. This causes cavitation to occur in the line to the pump, which is damaging to the pump. The pilot-operated check valve prevents this occuring as the valve will close to stop the load from moving.

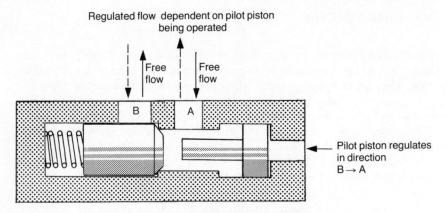

Fig. 2.2   Pilot-operated check valve.

2.4 THROTTLE VALVES

Throttle valves (Fig. 2.3) have a restriction or throttling device placed in a normally open orifice which limits the flow of fluid. There are different types of restriction.

1. ORIFICE: the restriction of flow through the body of this type of valve is caused by a small hole or jet. It is commonly used in the form of a screwed plug that has a hole drilled through it.

2. TAPER NEEDLE: oil is directed from the inlet to the outlet ports of the valve through a passageway of fixed size. A tapered plug is screwed into the passageway to restrict the opening. Oil is then controlled through the annulus space between the tapered plug and wall of the passageway.

3. WEDGE: this is similar to the taper needle but is used for very fine control. A cylindrical plug is inserted into a fixed orifice, this plug having a tapered flat manufactured on to it. Moving the plug out exposes a larger length of flat and gives a greater segment of area for the oil to flow through.

4. NOTCH: This throttling device uses a 'vee' notch cut into the length of the cylindrical plug. As the plug is withdrawn, a larger area of the triangular shape is exposed for the oil to flow through.

With all types of throttling device the pressure drops when passed through an orifice. The flow rate through the orifice is equal to the square root of the pressure drop multiplied by the area of the orifice times a constant. Increasing the pressure drop across the orifice will increase the flow.

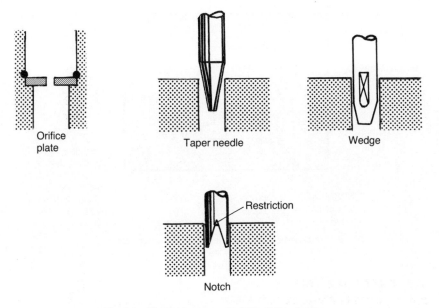

Orifice plate

Taper needle

Wedge

Restriction

Notch

Fig. 2.3   The various types of throttling device.

2.4.1   Pressure Compensation

Additions can be made to the valve to compensate for variation in the pressure or temperature. This is done by sensing the pressure on both the inlet and outlet sides of the valve and comparing them.

A spool is inserted which is capable of blocking the inlet flow to the throttling device. A spring is used to hold the spool in the open position. The inlet pressure is then allowed to act on the end of the spool which reacts against the spring. The pressure from the outlet side of the throttle is allowed to act on the spring end of the spool (Fig. 2.4).

As the pressure builds up, due to the pressure drop across the orifice, it moves the spool and closes off the inlet to the valve to match the flow which is passing through the restriction or orifice. Variations of pressure caused by the load are automatically compensated for in that the pressure which is due to the load acts on the spring side of the compensator spool. To maintain a constant flow at the outlet port when there is a variation of fluid viscosity the compensator spool must be constructed with sharp edges.

These valves are usually by-passed by means of a check valve, thus allowing free flow in one direction. They can be operated in one of four different ways:

1.   manually;
2.   mechanically;
3.   hydraulically;
4.   electrically.

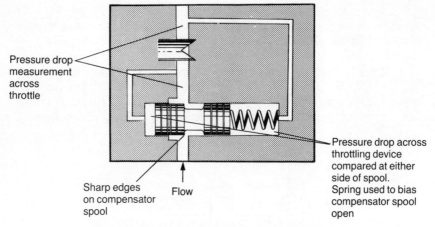

Pressure drop
measurement
across
throttle

Pressure drop across
throttling device
compared at either
side of spool.
Spring used to bias
compensator spool
open

Sharp edges
on compensator
spool

Flow

Fig. 2.4   **Pressure and temperature compensator.**

2.5 PRESSURE VALVES

Pressure valves are used to safeguard the components in a system from overload. Sometimes they are used to reduce the pressure in one part of the system. At other times they automatically sequence the movement of loads. It is possible to have pressure control valves working simultaneously in each of these ways in various parts of a hydraulic system.

There are three types of pressure control valve:

1. relief;
2. sequence;
3. pressure-reducing.

2.5.1 Relief Valves

Pressure relief valves are similar to check valves with the additional feature of allowing adjustments to the spring force. This in turn varies the pressure needed to open the valve. They are used primarily to limit the pressure in the system to protect the components from damage. They come in two design types:

1. direct-acting;
2. pilot-operated.

As the flow increases through a direct-acting pressure relief valve it causes the poppet to move further out of the seated orifice. Moving the poppet increases the compression to the regulator spring; this increases the pressure setting of the valve. Therefore, increased flow brings increased pressure.

With pilot-operated relief valves the movement of the poppet is very small, since the poppet is used for only a pilot function. The main poppet will open fully and in doing so passes a very small volume of oil across the pilot poppet.

Pilot-operated valves are used for relieving large volumes of oil through the main stage with little oil flowing through the pilot stage. They can also be fitted with a solenoid valve for unloading purposes. The solenoid valve directs the pilot oil back to the reservoir at the pressure of approximately 3 bar (45 psi).

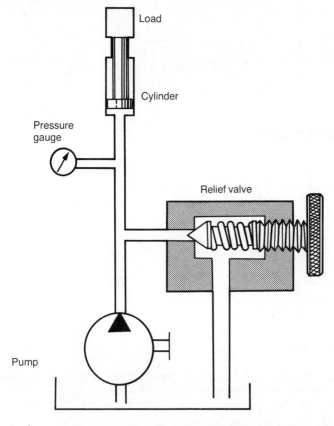

Fig. 2.5   Protecting a hydraulic system using a relief valve.

Figure 2.5 shows the effect that a direct-acting relief valve has in protecting the hydraulic system from overload. When oil is allowed to pass across a relief valve, the power which produces the flow at the given pressure is converted into heat. This heat is transferred to the reservoir by the flowing oil. It is always a good practice to unload the pressure generated by the pump from the hydraulic system when it is not being used to perform work. The aim is to unload the flow at as low a pressure as possible.

2.5.2 Sequence Valves

Sequence valves are very similar in construction to relief valves. However, their function differs: whereas relief valves spill the fluid to the reservoir, a sequence valve holds the fluid in the system. When the fluid pressure which is acting against the area of the poppet is greater than the spring force, the valve opens and passes the flowing oil to the next operation.

    Sequence valves are used to pressurize one part of the circuit before another part is allowed to operate. The sequencing is done automatically by the inlet pressure. As an example, a clamping and drilling function can be

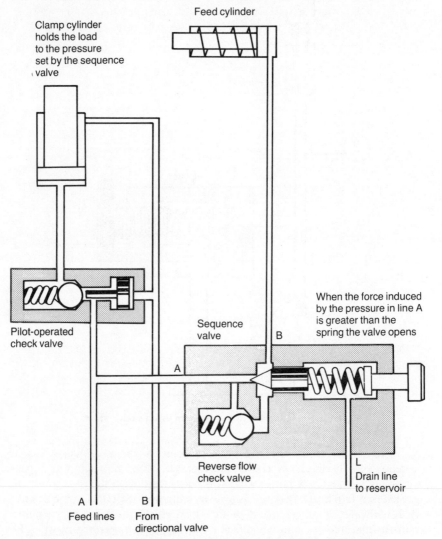

Fig. 2.6 Circuit diagram showing the sequencing of two cylinders.

done with a sequence valve. The operator would initiate the clamp function and when the pressure reached the predetermined force set by the sequence valve, the drill feed would start (Fig. 2.6).

To provide a reverse flow of oil a check valve is fitted to the circuit to allow retraction of the cylinders.

2.5.3 Pressure-reducing Valves

Relief valves will always limit the system pressure to the lowest valve setting in the system. At times it is necessary to have various pressure settings within the one system. In this case a valve called a pressure-reducing valve is used (Fig. 2.7).

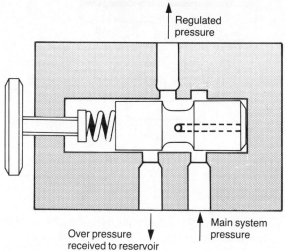

Fig. 2.7 Hydraulic pressure-reducing valve.

The principle of this valve is similar to the pressure compensator of the throttle valve (see section 2.4). Fluid is allowed to flow through the valve until the pressure induced by the load is great enough to close the valve against a spring force; it then closes and blocks the inlet to itself.

Some types of pressure-reducing valve have an extra port to return the oil to the reservoir. An increase in the pressure, induced by the load when the valve inlet is closed, relieves this pressure increase to the reservoir or to the atmosphere in the case of pneumatic valves. It is possible to exert the force on the spring with an electrical solenoid which will allow remote controlled operation. These are available as proportional pressure-reducing valves.

Pressure-reducing valves are sometimes called 'air regulators'. They are used extensively in pneumatic systems. They limit the higher receiver pressure to a level which is suitable for the various individual systems and the required load forces to the actuators. Figure 2.8 shows their design features.

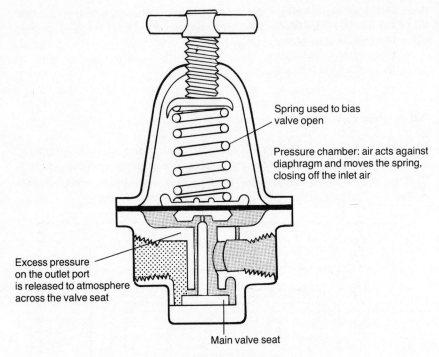

Spring used to bias valve open

Pressure chamber: air acts against diaphragm and moves the spring, closing off the inlet air

Excess pressure on the outlet port is released to atmosphere across the valve seat

Main valve seat

Fig. 2.8   Air regulator valve.

2.5.4   Pressure Switch

This valve is used in a hydraulic circuit to give an electrical signal when the system pressure reaches a desired setting. It is based upon two designs:

1.   piston;
2.   Bourdon tube.

In the piston design, pressure from the system acts against the piston and moves it against a spring. The pressure is adjusted with a screw acting against the spring. The travel of the piston activates a simple microswitch.

A Bourdon tube works on the principle of a curved hollow tube which attempts to straighten when pressure is applied within it.

The Bourdon tube type of pressure switch activates the limit switch as it straightens or flexes due to the system pressure. This type of valve is prone to signal fluctuations due to pressure transients.

2.6   DIRECTIONAL CONTROL VALVES

Directional control valves for hydraulics and pneumatics all follow the same principle and divide into five categories:

1. two position, one way;
2. two position, two way;
3. two position, three way;
4. two position, four way;
5. three position, four way.

One of the differences between hydraulic and pneumatic valves is that pneumatic valves have two outlets marked 'E' or 'R' (exhaust or return) which discharge to the atmosphere. Hydraulic valves have only one outlet, 'T' (tank), which returns oil to the reservoir.

Fluid which is at the inlet or pressure port (P) is delivered to either of the outlet or cylinder ports (A) or (B) by the movement of the spool (Fig. 2.9). It is usual to have either two or three position valves, which are shown in Fig. 2.10 as boxes, each showing the flow paths through the valve.

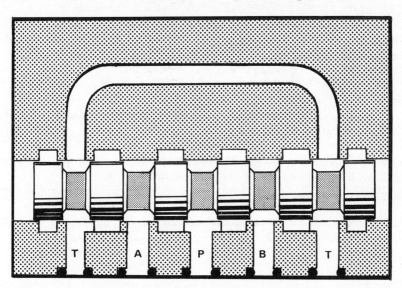

Fig. 2.9   Cross section of a hydraulic spool valve.

Three position valves have a center box which shows the condition of the valve at rest. There can be up to eight different options of valve design giving various features of control for fluid power systems.

To the layman, directional control valve designs seem to be very complex. A lot of mystique is implied when designers discuss the relative merits of a valve. However, it is important to remember that the body of a valve is produced in large quantities and is the same for every type of valve. The variation in the flow types is made to the valve spool. Note that all spool valves have a potential leakage path, some worse than others.

Spool valves are manufactured to very close tolerances with radial clearances between the spool and the valve body of 5–15 μm.

Alternative symbol for
center position

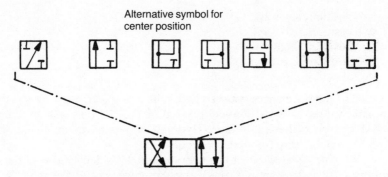

Fig. 2.10   Valve symbol shown as a series of boxes.

The edges of the spool lands may or may not cover the shoulders inside the valve body. These two states are called *overlap* and *underlap*.

Oil that passes through the internal clearances between the spool and valve body is called leakage (not to be confused with that which drips on the floor). The internal leakage of a valve is caused by a combination of radial clearance and the overlap of the spool in the valve body. Underlapped valves always have an internal leakage in the mid-position as all ports are interconnected.

Positive overlap is used when the pressure must not collapse during the switching operation. Caution must be taken with this type of spool as it causes operational shocks due to pressure peaks. It can be seen (Fig. 2.11) that as the spool moves to the left the flow between ports B and T is stopped before any flow between ports P and B is initiated.

Negative overlap momentarily connects all the passages together; the

During the transition state the pressure port is always blocked

Pictorial representation of valve overlap. This configuration of valve is usually manufactured as negative or underlapped.

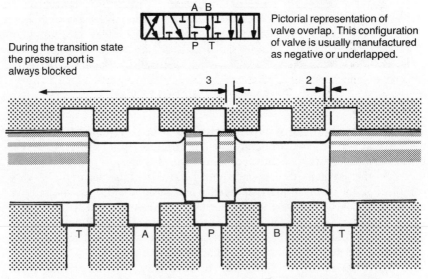

Fig. 2.11   Section of a valve showing overlapping of the spool and body.

pressure collapses and the load decays. This is illustrated (Fig. 2.12) by moving the spool to the right where the ports A, B, P and T are interconnected. However, transients with this type of valve are less aggressive than those with overlapped valves.

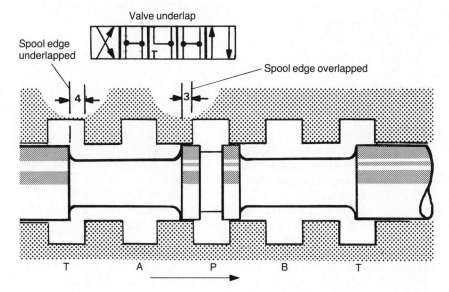

Fig. 2.12 Section of a valve showing the effect of underlap. As the spool moves in the direction of the arrow, the overlapped condition is cleared and all the ports are interconnected.

To ensure smooth acceleration of a load from rest, to control an actuator at very slow speeds or to position accurately, valve spools are produced with a taper or notches on the leading edge. These features allow for gradual increase in the flow as the spool moves from the set position. This reduces shock loading to the hydraulic system. The notches are either machined into the spool or produced by the action of a press (Fig. 2.13).

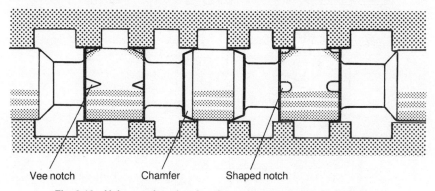

Fig. 2.13 Valve section showing the various types of spool edges.

2.7 ACTUATORS FOR OPERATING DIRECTIONAL CONTROL VALVES

To produce a flow across a valve the spool must be moved a set amount. To achieve this movement different actuators are used. The various types of actuator for moving the spool are:

1. manual;
2. manual with detent;
3. roller;
4. foot pedal;
5. hydraulic;
6. pneumatic;
7. electrical solenoid;
8. solenoid with spring return;
9. solenoid with spring centering;
10. solenoid and hydraulic pilot.

2.7.1 Manual Lever

The hand lever is usually attached to the spool by a drive shaft. Sometimes a spherical end is fitted into the drive shaft to give location to the hand lever. Movement of the hand lever moves the valve spool and creates flow from the pump port to the cylinder ports. Manual lever operators come with a detent which will hold the spool in an offset position, allowing the operator to perform another action. It is often used for mobile applications (Fig. 2.14).

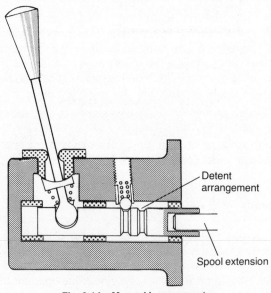

Fig. 2.14 Manual lever operator.

Manual lever valves are used so that the machine-operator can control the load and see how far it has moved.

2.7.2 Roller-operated

The roller-operated actuator consists of a plunger with a roller attached to it which presses against the valve spool. A mechanical action, possibly a cam driven by a cylinder, activates the plunger. This causes the plunger to move the spool against the internal spring (Fig. 2.15).

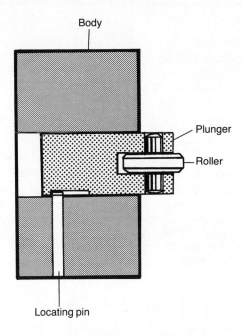

Fig. 2.15   Roller/plunger operator.

It is good practice to allow for an override movement of the plunger in order to save damage to the spool of the valve.

2.7.3 Foot Pedal

The foot pedal actuator (Fig. 2.16) is used mainly with pneumatic systems and gives a definite switching action, either on or off. It is used when both the hands of the machine-operator must be involved holding the component in position. It is not used if it is possible for the machine-operator to let go of the load with one hand.

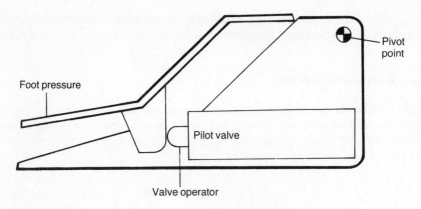

Fig. 2.16   Foot pedal operator.

2.7.4   Hydraulic Pilot

Hydraulic pilot control (Fig. 2.17) takes hydraulic pressure from a small pilot source and uses this to move a piston against the spool to produce the switching operation. Pressures as high as those set by the main relief valve can be used in some cases, but in the majority of instances the maximum pressure 70–100 bar (1000–1500 psi) is used for hydraulic systems and 4–6 bar (60–85 psi) is used for pneumatic systems.

When it is necessary to operate the hydraulic pilot piston at a lower pressure than the system pressure, and should a separate pilot pressure pump

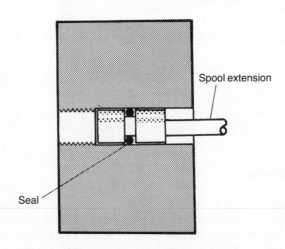

Fig. 2.17   The hydraulic pilot actuator.

not be fitted into the system, then a pressure-reducing valve is used to provide the pilot oil at a reduced pressure.

2.7.5 Pneumatic Pilot

Pneumatic pilot control (Fig. 2.18) works with low pressures of 7 bar maximum. A piston larger than that of the spool is required to move the spool against a combination of the spring and the internal fluid forces. The diameter of the pneumatic piston is sized to give a thrust against the valve spool which is much greater than the reacting spring force.

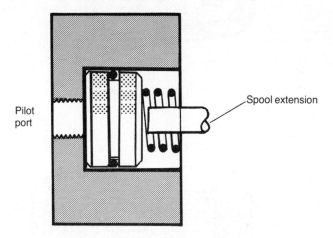

Pilot
port

Spool extension

Fig. 2.18   Pneumatic pilot actuator.

2.7.6 Electrical Solenoid Actuation

Electrical solenoids are manufactured for either AC or DC voltage; the two types are quite distinct in design.

1. AC solenoids (Fig. 2.19) have a high inrush current which, if allowed to operate for more than approximately 45 seconds, will result in failure of the coil, due to overheating. They are therefore designed with a T-stack armature so that the eddy currents can operate in a closed loop. This reduces the current drawn, saving the valve from damage. The main causes for failure are dirt in the system, causing sticking of the spool, or energizing the two solenoids simultaneously.

2. DC solenoids (Fig. 2.20) are more rugged in construction and will allow valves to be operated partially open. Generally, DC solenoids are recommended in favor of AC solenoids.

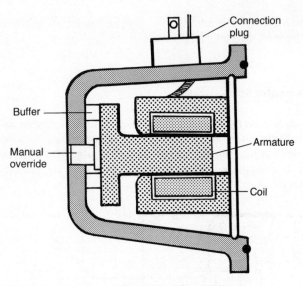

Fig. 2.19   AC solenoid.

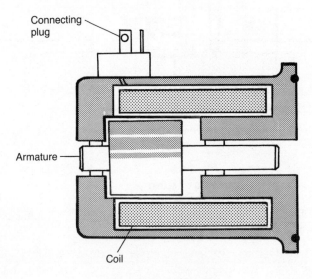

Fig. 2.20   DC solenoid.

2.7.7   Solenoid with Spring Centering

All direct-acting control valves and three position directional control valves use the spring centering method of actuation. The actuator is used for two position valves where the spring returns the spool to the rest position. Pump-unloading valves are usually two position spring return.

2.7.8 Solenoid/Pilot Actuator

Hydraulic valves used for larger flow rates and most pneumatic valves use the solenoid/pilot-operating principle. This allows solenoids with a low current force to be used to move spools which would require very large forces. The pilot section is a small directional valve which in its turn controls a larger directional valve (Fig. 2.21).

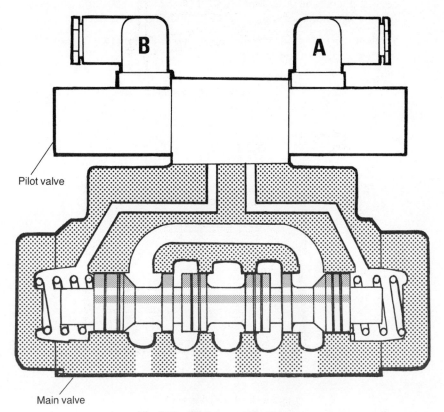

Fig. 2.21   Pilot-operated valve.

When flow rates exceed 100 l/min (22 imp. gals/min) it is customary to use a small valve as the pilot stage for a valve which has a larger flow rate. This type of actuator is used for valves with port sizes of up to 63 mm (2.5 in) diameter.

QUESTIONS

1. Name the main categories of valve.

2. Name the parameters that each category of valve controls.

3. When discussing flow control valves, what device is used to maintain constant flow regardless of the load?

4. Explain what is meant by overlap and underlap in valve design.

5. Each valve has its own descriptive symbol. Give the symbol for the following valves:
   (a)   a pressure-compensated flow control valve;
   (b)   a pneumatic directional control valve of three position, five way;
   (c)   a solenoid pilot control valve for hydraulic operation.

6. Explain what is meant by pilot operation and name two circumstances when it would be used.

3

Load Actuators

For a fluid under pressure to produce force and movement, it is necessary to have load actuators. This chapter describes the different types of actuator available to the engineer. Air-operated components will be described first, and then those operated by oil.

3.1 LOAD ACTUATOR TYPES

Load actuators are divided into three main types:
1. cylinders that give a linear movement;
2. motors that produce rotational movement;
3. rotary actuators that move through an arc.

All three components are readily available for use with air or oil.

3.2 AIR CYLINDERS

Air or pneumatic cylinders are capable of giving a thrust of up to 5000 newtons (1000 lbf). They come in two design types (Fig. 3.1):
1. single-acting spring return;
2. double-acting.

As the design types state, a single-acting cylinder is moved by the fluid in only one direction. Sometimes a large spring is fitted around the piston rod which is capable of returning the piston and load to the start position.

Double-acting cylinders are made with a fluid inlet port at each end of the cylinder barrel so that the fluid will move the cylinder in either direction.

There are various design features which can be added to the double-acting cylinder to allow it to perform different functions:
1. cushions, to decelerate the piston rod at the end of its stroke;
2. locking devices, to hold the load in a given position;
3. proximity devices, to indicate position at the limits of its stroke;

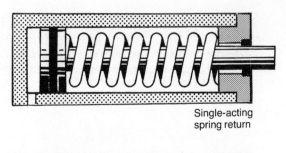

Single-acting
spring return

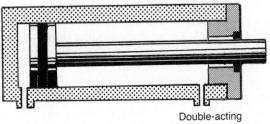

Double-acting

Fig. 3.1   The two types of cylinder.

4.   absolute position indicators, to indicate position along the length of its
      stroke;
5.   wide choice of cylinder mountings.

3.2.1   Cushions

Cushion adjustment is used to decelerate the piston rod as it reaches the limit
of its stroke. Care must always be taken to ensure that the deceleration forces
are not greater than the design strength of the cylinder.

    The cushion is usually a sleeve fitted around the position rod of the
cylinder, close to the piston. Its length is similar to that of the piston and it

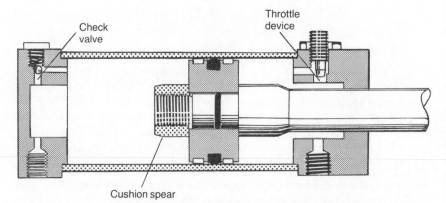

Fig. 3.2   Cross section of cylinder showing cushion controls.

slides into a cavity in the end-caps. This restricts the flow of fluid from the outlet port of the cylinder and reduces the momentum of the load (Fig. 3.2).

With some cylinder designs the effective area of the piston is reduced when the cushion enters the cavity. Intensification of pressure then takes place in the cushion cavity. This is caused by the input pressure to the cylinder and the deceleration forces of the load acting over the reduced area of the cushion.

3.2.2   Locking Devices

Locking devices (Fig. 3.3) are used mainly with pneumatic cylinders. These are devices which lock onto the piston rod and prevent movement of the load when the air is removed. The locking action is achieved by air acting against a piston which is attached to a locking cam. The eccentric force of the cam locks the cylinder piston rod against the wall of its bearing.

The piston rod is protected from damage in the locking action by the plastic material used in its construction.

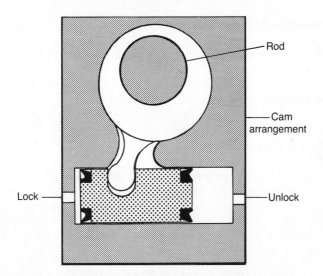

Fig. 3.3   Section of a typical locking device for air cylinders.

3.2.3   Proximity Devices

Proximity devices (Fig. 3.4) are magnetic devices which determine the presence of a metallic object. The magnetic field passes through the wall of the cylinder and the switching action takes place as the cylinder piston approaches. The switching action is produced by an electronic circuit housed in the body of the sensor, using the change of magnetic field.

Proximity devices act like built-in limit switches which indicate the presence of the piston at a set position. They are mounted on the outside of the cylinder and are usually adjustable. The cylinder tube is usually made from a non-ferrous material.

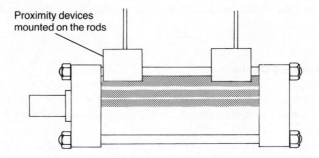

Proximity devices
mounted on the rods

Fig. 3.4   Cylinder showing proximity devices mounted on the tie rods.

3.2.4   Absolute Position Devices

As a progression from the proximity device, the absolute position sensor has been developed to provide a definite indication of the position of the piston rod in the cylinder; see section 5.4.3. Electrical values are given from the measuring device which is mounted inside the cylinder.

3.2.5   Cylinder Mountings

Tie rods are used in the designs for the majority of air cylinders. This is to allow ease of manufacture for the various lengths of stroke. The smaller bore sizes of cylinder come with the end-caps screwed to the cylinder barrel.

There are numerous methods of mounting a cylinder by fastening different mounting attachments to the cylinder body and end-caps. The types of mounting available are (Fig. 3.5):

1.   flange mounted, with the flange mounted on either the front or rear end-cap;
2.   foot mounting, with the foot attachment mounted on both the end-caps;
3.   clevis mounted, with the clevis attachment fitted to the rear end-cap;
4.   trunnion mounted, the trunnions are specially manufactured end-caps or an extra attachment fitted along the length of the cylinder barrel.

The rods usually have a clevis mounting from a swivelling bearing or a fixed fork end. These are attached to a screw thread produced on the cylinder piston rod.

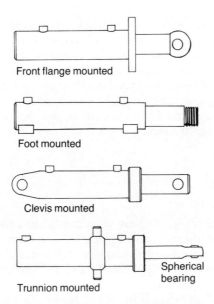

Front flange mounted

Foot mounted

Clevis mounted

Trunnion mounted

Spherical
bearing

Fig. 3.5   Different types of cylinder mounting.

3.3 HYDRAULIC CYLINDERS

Hydraulic cylinders convert the pressure within a hydraulic system into force. Hydraulic cylinders need to be stronger than air cylinders because of the increased forces involved.

It is customary for each of the design types to be used by a particular branch of industry:

1. Tie rod—used mostly in machine tools where the pressures involved are lower;
2. welded—used mostly by the construction industry where low costs and robustness are required;
2. flanged—used mostly by heavy industry where the cylinder acts in some cases as a strut;
4. displacement—generally used for single-acting applications where the load returns the extended piston rod.

Spring return cylinders are sometimes used but in the main they are not acceptable.

3.4 RODLESS CYLINDERS

The rodless cylinder is a development which allows the load to move within the overall length of the cylinder. A steel band is attached to both ends of the internal piston and sealed by the end-caps. The load mounting is then attached to the steel band and moves as the internal piston moves.

To move a load from left to right, compressed air is fed to the right-hand chamber and the steel band pulls the load along the side of the barrel; the steel band is then in tension and no buckling takes place.

A rodless cylinder is very similar to a linear encoder and in fact it may be possible to combine these components to give position control to the accuracy of the encoder (see section 5.4.2).

3.5 AIR MOTORS

Vane type air motors (Fig. 3.6) are very similar in construction to vane pumps but operate by compressed air being fed into them. They are of fixed capacity and give a specific torque output in relationship to the pressure of the air. The vane acts as an air paddle and the movement of the vane produces a rotation of the shaft.

Another type of air motor is the piston motor. This is constructed with the pistons attached to a crankshaft and the pistons move radially in the body

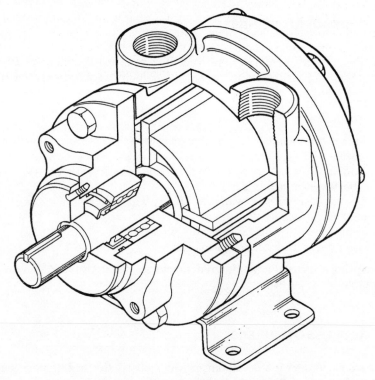

Fig. 3.6   Section showing vane motor.

or housing. To achieve high torques a gear ratio must be applied to the output shaft. These motors are high speed, low torque drives.

Air motors are not damaged if overloaded, unlike electrical motors which burn out.

3.6 HYDRAULIC MOTORS

Hydraulic motors are used to provide rotary movement from a hydraulic force and will rotate at speeds of 5–6000 r.p.m. Hydraulic motors, like pumps, can be built in four different ways:

1. vane;
2. piston;
3. gear;
4. orbiting gear principle.

3.6.1 Fixed Displacement Motors for High Speed

Vane motors are very similar in construction to fixed vane pumps apart from the inlet and outlet ports (see section 1.2). Gear motors are similar in construction to external gear pumps, apart from reversible motors.

Reversible motors have a small modification which allows the fluid, under pressure, to enter either port and provide a forward and reverse drive motion. The internal sealing of these motors is different from that of their respective pumps.

Vane and gear motors are usually high speed motors with minimum speeds of 300–500 r.p.m. They can achieve maximum speeds of 6000 r.p.m. but at low torque outputs.

3.6.2 Piston Motors

Hydraulic piston motors are produced in the two design types shown below. Each type of motor is used for a specific application:

1. radial piston, used for high torque, low speed applications;
2. axial piston, used for high speed, low torque applications.

Radial piston motors are designed in such a way as to give good starting and stalled torque capabilities and are suitable for applications where the drive starts and stops under load. The starting torque varies with the cam shaft angle; Figure 3.7 shows the maximum and minimum values.

The effect of viscosity is negligible but back pressure in the outlet port will produce a reduction in torque output. This fact has to be considered when using two motors together in series.

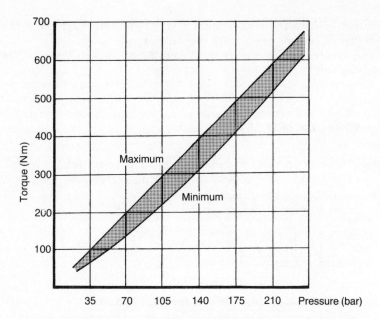

Fig. 3.7   **Graph showing maximum and minimum starting torques.**

Speed of operation affects the volumetric efficiency of a motor whatever the design. If a motor is stalled under pressure, the oil will leak away through the manufacturing clearances and produce losses which, in turn, will be converted into heat.

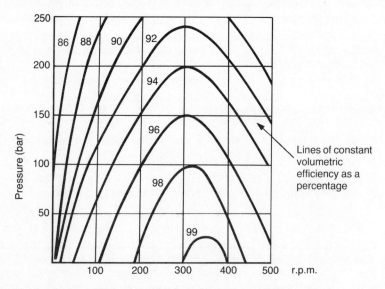

Fig. 3.8   **Graph showing the hydraulic efficiency curves as a product of pressure and speed.**

Figure 3.8 shows the volumetric efficiency which is typical for a radial piston motor.

As can be seen, the leakage is least when the speed is between 300 and 400 r.p.m. and the pressure is less than 35 bar.

The overall efficiency of a radial piston motor is dependent on many factors such as torque, speed, pressure and friction. Figure 3.9 shows the overall efficiency of the motor by comparing the output power with the input power. The maximum efficiency occurs between 200 and 300 r.p.m. and at a pressure of 150 bar (200 psi). Obviously volumetric efficiency has been sacrificed for better overall efficiency.

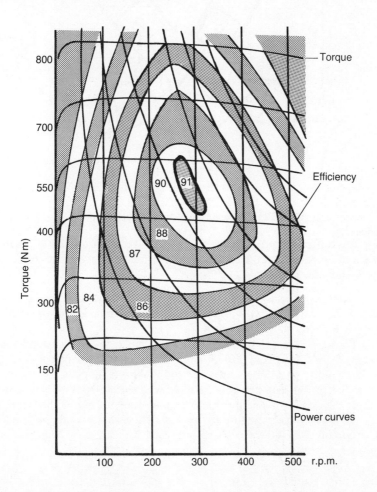

Fig. 3.9 Graph showing torque, power and overall efficiency of radial piston motors.

The motor is designed with a number of radial pistons; usually an odd number, typically five. The thrust from the piston under pressure is taken at an angle to the crank shaft. As the shaft rotates, the load from the first piston decreases. The second piston starts to exert a force on the crankshaft and takes over the main thrust, acting on the shaft. Each piston in turns exerts a radial thrust on the shaft to produce a turning motion as shown (Fig. 3.10). The control logic for giving the maximum pressure at the right cam angle is manufactured into the rotating port arrangement and body of the motor.

The effect that the different pistons have is shown in Fig. 3.10.

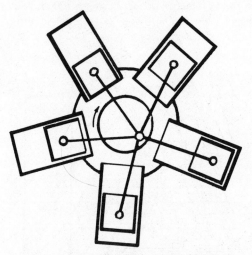

Fig. 3.10   Section of a typical radial piston motor.

It is difficult to produce a variable speed, radial piston motor because of the variation needed in the cam angle. An attempt to overcome this is made by switching in another bank of pistons.

Axial piston motors (Fig. 3.11) are virtually the same as axial piston pumps (see section 1.2.2). Variable speed motors are available but vary only by approximately 50% of the speed.

Torques from axial piston motors are much lower than radial piston motors. Typically an axial piston motor will develop 75 kW (100 h.p.) at 6250 r.p.m. A radial piston motor would produce the same power at 150 r.p.m. The radial piston motor in this example would develop 4300 Nm (860 lbft) whereas the axial piston motor would only produce 143 Nm (35 lb ft) torque.

To provide the same torque the axial piston motor would be fitted with a gearbox of 30:1 ratio. Gearboxes are used frequently for winches, wheel and track drives.

3.6.3   Orbiting Gear Motors

The orbiting gear motor generates a hypercycloidal motion as it rotates. The rotor has a number of lobes manufactured on it with gear teeth machined on

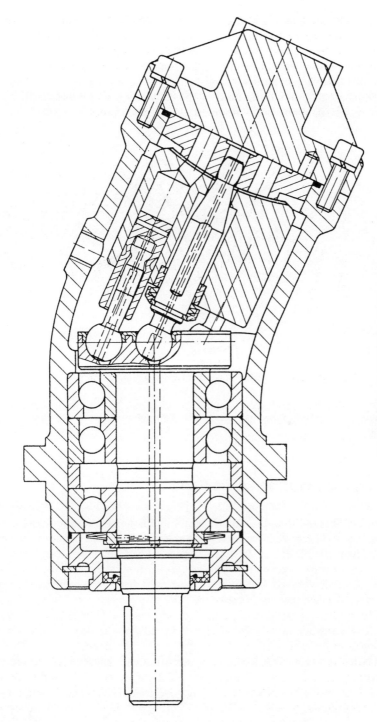

Fig. 3.11   Bent axis piston motor.

the inside. The rotor moves under the influence of the fluid, around the fixed rim which has mating lobes on it.

Fluid under pressure, shown as the shaded area in Fig. 3.12, moves the rotor through one pitch of the outer rim. The drive shaft rotates in the same direction by 1/42 of a revolution.

The logic to provide the fluid in the correct location is produced by a plate valve, having many slots, mounted into the end covers.

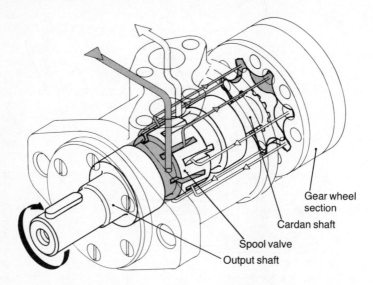

Gear wheel
section

Cardan shaft

Spool valve

Output shaft

Fig. 3.12   Section showing the principle of operation of an orbiting motor.

3.7 HYDROSTATIC TRANSMISSIONS

It is not possible to discuss hydraulic motors without mentioning hydrostatic transmissions. These are integrated drive systems of a pump combined with a motor. The principle of this drive is to provide a high torque output at a slow speed and maximum speed at low torque. Maximum power is achieved from the motor at slow speed and so it is ideal for wheel drive applications on vehicles for the construction industry.

Hydrostatic transmission (Fig. 3.13) provides a drive which is reversible. It will reverse at maximum speed, causing the minimum amount of mechanical shock to the system.

Hydrostatic transmission circuits are fitted with relief valves to reduce the effect of pressure peaks. These valves allow the by-pass of total flow of the pump or motor and act as a safeguard against pressure damage through high pressure transients.

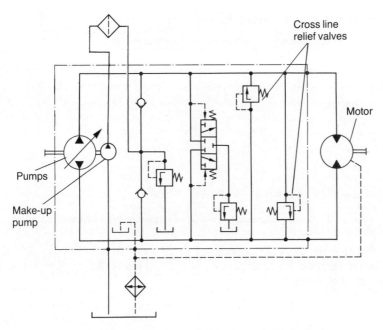

Fig. 3.13   Circuit showing typical hydrostatic transmission.

The system is piped up as a complete loop. The flow of oil from the pump is totally used by the motor and returns to the pump again. To reduce the build-up of heat by the loss of oil through leakage, an additional pump is normally fitted to the transmission pump. It is designed to replace 25% of the total oil flow in the closed loop transmission with oil from the reservoir which has been filtered of dirt particles and cooled. It is not uncommon for 20% of the input power to be converted into heat with hydrostatic transmission systems.

3.8   ROTARY ACTUATORS

Rotary actuators (Fig. 3.14) are produced with various arcs of movement such as 45°, 90°, 180° and 360°. They are made as piston and vane types.

These devices are capable of moving a load through an arc. Rotary actuators capable of exerting high torque forces are usually operated through a rack and pinion drive. Seals are fitted to the ends of the rack to produce a fluid-tight chamber similar to that of cylinders. Fluid enters this chamber and moves the rack, producing a rotary movement.

The vane type of rotary actuator is very difficult to seal and therefore is limited in its application. The normal application for vane type rotary actuators is the 'wrist' movement for loading arms.

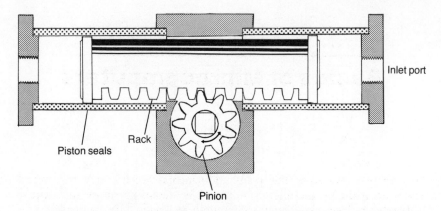

Inlet port

Piston seals

Rack

Pinion

Fig. 3.14   Section showing the action of a cylinder-type rotary actuator.

Piston type rotary actuators when fitted with check valves can hold the load in set position for long periods of time. This is due to their intrinsic positive sealing action.

QUESTIONS

1.   Name the different types of load actuator.

2.   What is the difference between a single-acting and a double-acting air cylinder?

3.   To decelerate the cylinder piston-rod at the end of its stroke what component is used and what care should be taken?

4.   Why are locking devices required for air cylinders?

5.   What type of actuator would be used for moving a load through 60 degrees of arc?

6.   In a pneumatic system what would happen if the air motor was temporarily over-loaded?

7.   Describe the different types of hydraulic motor.

8.   What protection is required in a circuit for a hydraulic drive to prevent damage from over-loading?

4

Basics of Microcomputers

This chapter describes the main components which, with the microprocessor, make up the microcomputer. The details given are sufficient to enable the engineer to use microcomputers effectively. This chapter discusses the structure of the microcomputer, including memory, the means of connecting to external devices, and power supplies. Programmable controllers are also briefly considered.

4.1 THE MICROPROCESSOR

The word 'microprocessor' is popularly used when 'microcomputer' is meant, in much the same way that 'steam engine' is used for 'steam locomotive'. To be exact, a microcoprocessor forms only part of a microcomputer. Microcomputers are based upon digital techniques, are very small and are inexpensive enough to be used in limitless engineering applications.

Microcomputers have to be programmed to operate in a required sequence. The fluid power engineer is usually the one who is most familiar with the control sequence of the equipment, so the programming is best done by him.

4.2 THE MICROCOMPUTER

The microprocessor is really quite useless on its own. It must be used with other devices. A typical arrangement which becomes a microcomputer is shown in Fig. 4.1. The basic elements are: microprocessor, to perform logic operations; memory, to hold programs and data; input/output device, to connect the external devices to the microcomputer; and a power supply.

Such a system can be on a single board which comes complete, ready made and tested. The engineer has to interface the microcomputer with the fluid power system and provide a power supply and program, (see chapters 6 and 8). Other devices added to the basic microcomputer are known as peripherals and well-known examples are printers and visual display units.

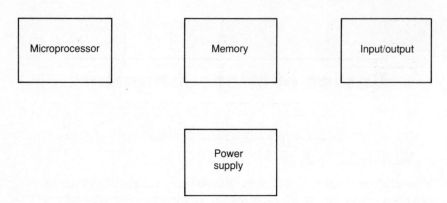

Fig. 4.1   Basic microcomputer elements.

4.3   MODE OF OPERATION

The microprocessor is the 'brain' of a microcomputer. It does its work by moving data or information between itself, the memory and the input/output device. It performs simple logical operations on data. The sequence in which it performs these operations is stored in memory in the form of a program.

Data are coded in binary form or in terms of binary digits. These are expressed mathematically in either 1s or 0s and called bits. Information is handled in units of a fixed number of bits. For example, an 8-bit microprocessor moves data in 'bytes' or 'words' or 8 bits, whereas a 16-bit microprocessor handles 16-bit words (see chapter 8).

The rate at which a microprocessor executes instructions is governed by a timing device. This timing device is often referred to as the 'clock' and is usually a quartz crystal. Typically one instruction is executed in 1 millionth of a second (1 μs).

4.4   THE BUS

The interconnecting structure of the microcomputer is called the 'bus' (Fig. 4.2). An 8-bit microprocessor will have 8 lines for transferring words of data, perhaps 16 lines for carrying a 'memory address', and a few lines for control and power supply.

Data are transferred through the bus system to the various devices. Information on locations of addresses are transferred via the address bus.

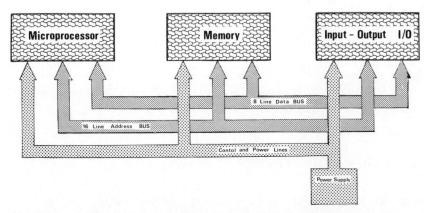

Fig. 4.2   Bus structure showing the interaction between the components.

4.5   MEMORY

The memory of a microcomputer holds the program or programs currently being used and provides storage for information such as input values, results of calculations, etc. The engineer must consider two types of memory as these will invariably be present in all systems.

4.5.1   Read Only Memory

Read only memory (ROM) comes in several types:

1. Mask programmed ROM in which the contents are 'manufactured in', giving a high initial cost but low unit cost. Hence it is used for high volume products.
2. Programmable ROM (PROM) in which the user can determine the contents by burning out selected links in the circuit. It is used for medium volume products.
3. Erasable PROM (EPROM) in which the user inserts the contents electrically (using an EPROM programmer) and can erase the contents by means of ultraviolet light. It is used in low volume and development applications.
4. Electrically erasable PROM (EEROM). This holds its contents until erased by an electrical signal.

4.5.2   Random Access Memory

In random access memory (RAM) or read/write memory the contents can be 'written into' and 'read from' by the microprocessor. It is used for temporary storage of programs and data since its contents are lost when the power supply is removed. It is known as volatile RAM.

RAM can come with a battery back-up. It will keep its contents for several years and in effect behave like ROM. It is then known as non-volatile RAM.

Where possible, permanent programs should be in ROM because ROM cannot be corrupted by electrical interference (see chapter 7).

4.6 OTHER MEMORY DEVICES

Generally microcomputers used to control fluid power equipment are 'dedicated' systems and will only require ROM (usually EPROM) and RAM. Other means of storage are:

1. Discs: these are used when a lot of memory is required. The engineer may come across them when producing programs on a 'development system' with discs. A development system is essentially a general purpose microcomputer with extra facilities. Disc drives are not very robust and are unsuitable for use in harsh environments.
2. Tapes: programs can be stored by tape recorder and transferred into RAM for execution. This is a popular way of storage for general purpose microcomputers but is complicated and not always reliable. It is widely used for home computers.

4.7 INPUT/OUTPUT

The input/output (I/O) of a microcomputer (Fig. 4.3) is the part which is accessible for connecting to external devices. It will usually require an interface (see chapter 6). There are various methods of obtaining input/output: parallel output, direct connection to the data bus, isolated input/outputs, and serial input/output.

4.7.1 Parallel Port Device

This is a component which provides ports or connections into or out of the microprocessor. Each port is usually of eight bits each in parallel; this is due to the configuration of the components which make up the I/O devices (Fig. 4.4).

A port can be treated as a whole word or a series of single bits. In some cases each bit can be used as either an input or an output as required. An example of an input would be a signal from a limit switch, and for an output the signal to control the energizing of a solenoid on a valve.

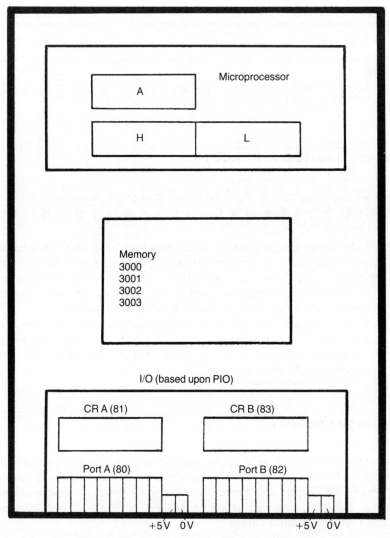

Fig. 4.3 Microcomputer showing the input/output.

Such devices are very versatile, easy to apply and widely used in engineering applications.

Well-known chips used for I/O are:

1. PIO, parallel input/output;
2. VIA, versatile interface adaptor;
3. PIA, peripheral interface adaptor;
4. RAM I/O, RAM input/output.

The Z80 microprocessor uses the PIO and therefore all applications in this book are based on it. Details of the PIO are given in appendix D.

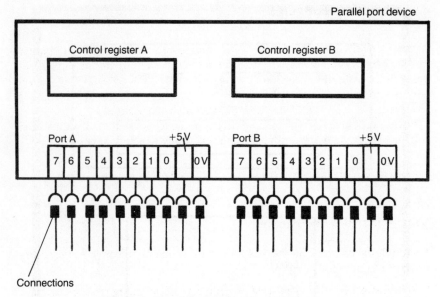

Fig. 4.4   Pictorial representation of a port.

4.7.2.   Direct Connecting to the Bus

The microcomputer can be expanded by connecting components to the bus. Tailor-made plug-in units, say a board to switch eight solenoids, can be obtained.

4.7.3   Isolated Input/Output

Some microprocessors, e.g. Z80, have locations reserved for I/O with a special set of instructions for fast access and simplified programming. The locations are not part of the normal memory.

4.7.4   Serial Port

Data are transmitted by a series of timed pulses rather like Morse code, but the receiving unit has to have a way of understanding what the pulses mean. It saves having a mass of wires as one wire will suffice. The rate at which these pulses are transmitted is called the 'Baud' rate. It would only be used where the fluid power system was some distance away from the microcomputer, because of the need to have devices to interpret the signals.

4.8 DC POWER SUPPLIES

A complete fluid power system with microcomputer control will require various power supplies at different voltage levels. The required stability of the voltage will differ for the various elements. For example, a typical microcomputer needs a stable supply at 5 V whilst a solenoid will have a moderately stable 24 V DC supply or even an AC voltage supply.

Power provided by the mains needs to be converted to produce DC voltage. There are two main types of power supply which do this:

1.   linear (low frequency);
2.   switch mode (high frequency).

4.8.1 Linear Power Supplies

The circuit shown in Fig. 4.5 has four main elements to convert alternating current into smooth DC supply:

1.   transformer, which is used to convert AC at mains voltage of, say, 240 V, 50 Hz to the required voltage at 50 Hz;
2.   rectifier, which is used to convert the transformed voltage into a fluctuating DC voltage;
3.   capacitor, which is used to remove the fluctuations in the DC voltage;
4.   Zener diode, which may be added to provide a standard voltage.

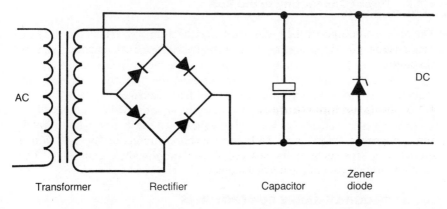

Fig. 4.5   Simple DC power supply showing the components.

A linear power supply will be:

1.   inexpensive;
2.   easy and quick to construct using the simple components shown. It may be purpose-built to suit any particular requirements.

The drawbacks for such a system will be that:

1.   the transformer is large and heavy;
2.   a large capacitor is needed;
3.   the output can be affected by mains interference;
4.   the low efficiency of about 30% means that a lot of energy is wasted as heat.

An example of the use for such a power supply is the powering of solenoid valves.

4.8.2 Switch Mode Supplies

The total system is sophisticated. In simple terms, the method of operation is that the input power is first rectified and then switched on and off at about 20 kHz. In effect, a high frequency source is produced, which can then be harnessed in much the same way as the linear power supply.

The characteristics of this type of power supply are that:

1.   it is small and light;
2.   it gives very little power loss and hence low heat generation;
3.   it operates from a widely fluctuating input voltage.

The switch mode power supply is frequently used for powering microcomputers.

4.9 BUFFERS

Microcomputer input/outputs are normally buffered. A buffer is a small amplifier which is considered as being expendable. When excessive power surges are experienced by the microcomputer, the buffer is destroyed, thus isolating and protecting the microcomputer from damage.

The term 'buffer' is also used to describe an intermediate storage file for data in memory. This is not something with which the engineer will normally be involved.

4.10 PROGRAMMABLE CONTROLLERS

4.10.1 General Description

Modern programmable controllers are no more than microcomputers used for control purposes.

The use of the name 'programmable controller (PC), or more archaically, 'programmable logic controller' (PLC), arises because programmable controllers, albeit very unsophisticated logic devices, were available before the microcomputer came on the scene. The retention of the name is more acceptable psychologically to engineers already versed in logic control.

Previous devices used binary switching, just as microcomputers do. They incorporated such components as relays, solid state (transistor) switches or fluidics (air flow logic). They were programmed to use the minimum number of components. Their performance was limited by speed of operation, sheer bulk and cost of components.

A microprocessor may contain 100 000 switches. Memory for programs is inexpensive and plentiful. The limitations of the earlier programmable controllers, therefore, disappear. Programs are now best written for ease of understanding, rather than to minimize components, and one may be as profligate with memory as one chooses.

4.10.2 Facilities

Programmable controllers will usually have the following facilities.

1. Control interfaces which are readily available either integral with the basic unit or as plug in extras. These interfaces are described in chapter 6.
2. Some means of permanent storage of developed programs which will be either RAM with battery or PROM.
3. Interaction with humans is often by a simple numeric keypad. Read-out may be on a simple display of light-emitting diodes. A full keyboard and monitor can be used if appropriate.
4. They are robust in construction and well-protected against electrical noise and interference (see chapter 7).

4.10.3 Relay Ladder Logic

Programming may be done in BASIC or machine code, as described in chapter 8, but often programmable controllers are designed to use *relay ladder logic*.

Relay ladder logic, as its name suggests, is based upon relay logic using boolean expressions. Proprietary PCs each have their own set of instructions, and programming manuals for individual PCs need to be consulted.

Just as a program written in BASIC has to be converted into machine code by a special program called an interpreter, so relay ladder logic has to have an interpreter. Relay ladder logic is:

1. as hard to learn as machine code;
2. slower than machine code in execution;
3. much more limited than machine code or BASIC.

It is suggested therefore that any engineer coming new to the subject should choose systems using BASIC and machine code rather than relay ladder logic.

QUESTIONS

1. What is the difference between a microprocessor and a microcomputer?

2. What are a 'bit', a 'byte' and a 'bus'?

3. List the different types of memory.

4. What is a port?

5. List the different types of input/output device.

6. Explain the difference between linear and switch mode power supplies.

7. What is a programmable controller?

5

Feedback Elements

So that a system can be properly controlled it is necessary to have means of monitoring its behavior. The measuring devices most commonly used in fluid power applications are covered in this chapter.

5.1 INTRODUCTION

Feedback is the signal from a monitoring device which indicates how a system is behaving. With a feedback signal the position of, say, a cylinder or the speed of a motor can be ascertained, and an input or control signal adjusted accordingly. Feedback has been used in this way for many years.

Microcomputers can perform this control function with the added advantage that logic can be included. Take as an example a traditional analog system controlling a cylinder, using a servo valve with an analog feedback device; it would not be able to correct the problem or halt the process if the cylinder was stopped part way along its stroke; in fact it would try to increase the flow from the valve to the cylinder. If a microcomputer were used, it could determine the stopped condition and could close down the hydraulic system.

Depending upon whether feedback is present or not, control systems are classified as either open loop or closed loop.

5.1.1 Open Loop

Systems without feedback are said to be 'open loop'. In the example shown in Fig. 5.1, when the microcomputer sends a signal to move the piston there is no way of checking that this actually happens. There is thus no means for the microcomputer to determine whether the cylinder has reached the end of its stroke.

5.1.2 Closed Loop

Systems with feedback are said to be 'closed loop'. In the example shown, Fig. 5.2, contact with the limit switch provides a feedback signal which verifies to the microcomputer that the cylinder is actually out.

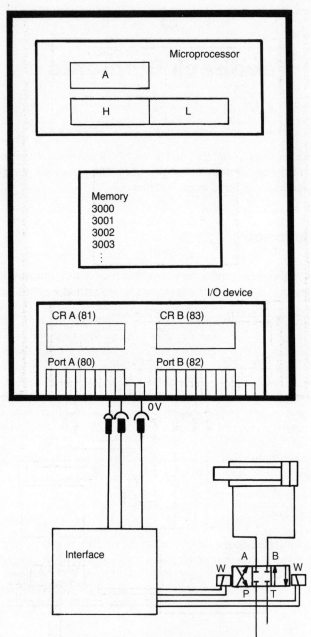

Fig. 5.1 A control system without feedback.

A simple limit switch only verifies one position. It would be better if the monitoring of position were continuous since then position, speed and velocity could be determined.

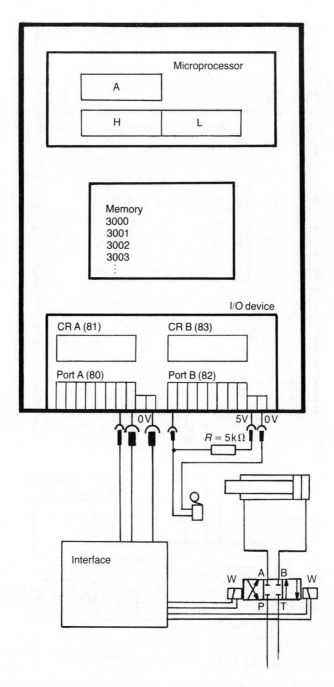

Fig. 5.2   The same control system as in Fig. 5.1, but the end of the stroke is verified.

5.2. CONTROL OF POSITION

Generally in fluid power applications the requirement is to be able to sense position. A microcomputer can determine velocities and accelerations from the time taken to move between positions.

Sensors for measuring position, as well as pressure and load are either digital or analog (see section 6.4). Usually with microcomputers it is easier to use digital devices.

5.3 PUSH BUTTON AND LIMIT SWITCHES

Push button switches (Fig. 5.3) are mechanical devices which operate by the closing of contacts, and are familiar to all. They are inexpensive and are widely and easily used for giving information to a microcomputer. Such switches could be a machine operator's 'start' and 'stop' push buttons.

Limit switches (Fig. 5.3) are mechanical switches which are activated when physically touched by, say, the movement of a cylinder piston rod. The drawback is that a limit switch can only provide information about a single position.

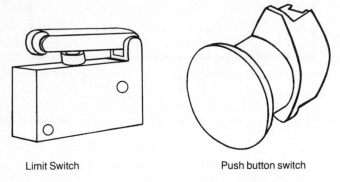

Limit Switch   Push button switch

Fig. 5.3   Push button and limit switches.

Mechanical contacts bounce, and a switch may appear to have been pressed several times when in fact it has only been pressed once. The programmer must allow for this when writing the program (see section 7.3).

5.4 DIGITAL DEVICES

Digital devices provide signals which give information in the form of a binary sequence. The information is by the pattern and sequence of pulses rather than the levels. Morse code is a good example of a binary sequence in that the different letters are represented by a combination of long and short pulses.

Some common digital devices available are light-activated switches, proximity sensors, shaft encoders, and digital cameras.

5.4.1 Light-activated Switches

Light-activated switches (LAS) use photosensitive diodes (Fig. 5.4) and are configured so that when a beam of light falls on them, a signal is given out, typically 5 V. When the light beam is interrupted, the signal changes, typically to 0 V.

The features of light-activated switches are:

1. they are inexpensive, robust and very reliable;
2. they provide a noncontact means of measurement;
3. they form the basis of many other sensing devices;
4. they need a simple supporting electronics circuit, usually with an adjustable potentiometer, to alter the switching level to match the light intensity.

Many digital measuring devices are based on the light-activated switch.

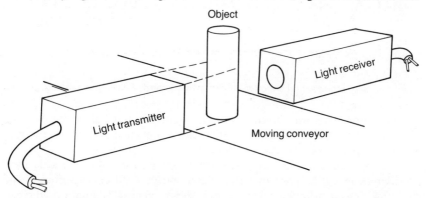

Fig. 5.4   Simple light-activated switch used for, say, counting components.

5.4.2 Optical Shaft Encoders

Optical shaft encoders are devices commonly used with microcomputers to monitor shaft displacement or position. They provide output signals which can be fed directly into a microcomputer typically via the input/output port. By counting the number of pulses with respect to time, velocity and acceleration can also be obtained.

The principle of operation is that light shining on a sensor, perhaps a light-activated switch, is interrupted by the bars, between a row of slots in a rotating disc. The voltage output corresponds to the slots as they pass the light.

A slotted disc can be made from metal with holes cut in it, but usually it is made of glass or plastic with a pattern etched upon it by a photographic process. There can be as many as 2500 slots positioned in a series of rows around the disc.

Incremental Encoders

Incremental encoders (Fig. 5.5) are based upon a disc with a single row of uniformly spaced slots. The switching signal is conditioned to give a square wave output. Relative distances are determined by counting the number of pulses generated as the shaft rotates and relating these to the circumference of the wheel or roller.

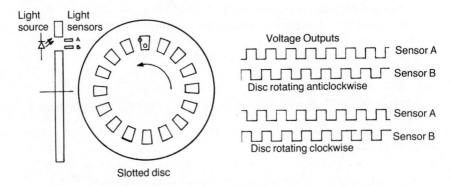

Fig. 5.5   Incremental encoder showing the method of direction sensing.

Usually there are outputs from two sensors set at half a slot distance apart so that (a) the direction of rotation can be detected and (b) the number of readings per revolution can be doubled.

If the diagram, showing the output voltage from the two sensors is considered, it can be seen that the relative outputs differ according to the direction. Sensor 'A' going from low to high with sensor 'B' high would mean anticlockwise rotation, and with sensor 'B' low would mean clockwise rotation. Also if the positions of the edges of the signals are considered, there are four distinct positions per slot. For example, a device with 100 slots can give 400 position changes per revolution.

A third output may also be provided to give a reference pulse once per revolution.

NOTE: vibrations can cause the shaft to move to and fro, thus generating pulses as if continous movement were taking place. When used with a microcomputer this can be corrected by programming.

Enhancements are available which effectively subdivide the width of each pulse in order to measure very fine parts of a revolution. They require sophisticated electronics to analyze the signals produced and are no longer simple digital devices.

One method of enhancement is to utilize the voltage directly from the light sensor, before it is made into a square wave. This will be roughly sinusoidal in output and hence have different values for the different positions of movement within the pulse.

Another method of resolution enhancement is to use Moiré fringes. This method is based upon two sets of gratings (rows of narrow slots) set at a slight angle to one another. Small relative displacements can be detected because they will cause the fringe, the interference between the lines on the gratings, to move substantially. The effect can be demonstrated easily by moving two ordinary hair combs over one another, at a slight angle to each other.

Absolute Encoders

Absolute encoders use several concentric rows of slots and so absolute position can be obtained; for example, 4 rows of slots will give 16 different positions, 5 rows 32 different positions. The maximum number of rows that can be achieved is about 10, and will give 1024 positions.

If a binary code were used, gross errors would result as there are positions where several bits change at the same time, for example in the change between 7 and 8. In a less than ideal situation these change-overs will not be exact, and there will be intermediate states where some bits have changed over and some have not. Readings taken at these conditions will be grossly in error. The usual way of overcoming this problem is to use what is known as a Gray-coded device (Fig. 5.6). There is still, of course, the problem of equating the readings with a more familiar numbering system. This can be done by electronic logic circuitry or by programming.

Mathematically, the Gray code is generated from the binary code by leaving the most significant bit, then changing the next bit if the preceding bit is 1 and so on.

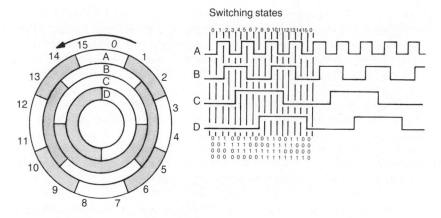

Switching states

Fig. 5.6   **Example showing the outputs from a 16 position Gray-coded device.**

Decimal	Binary	Gray code
0	0000	0000
1	0001	0001
2	0010	0011
3	0011	0010
4	0100	0110
5	0101	0111
6	0110	0101
7	0111	0100
8	1000	1100
9	1001	1101
10	1010	1111
11	1011	1110
12	1100	1010
13	1101	1011
14	1110	1001
15	1111	1000

Linear Encoders

Measurement of linear displacement can be achieved by simply connecting the shaft directly to the movement. True linear devices exist where the cursor slides over a linear scale. The same principles as for rotating encoders apply. Scale length is a limitation, and linear encoders are easily affected by dust and dirt. Rotating encoders are best used whenever possible.

5.4.3 Proximity Sensors

Proximity sensors sense magnetic material (Fig. 5.7) and a voltage is given out when the target is within a certain distance. They are commonly used in machinery with a rotary toothed wheel. As each tooth passes the sensor a square pulse is generated; hence the number of teeth passing the sensor can be monitored. A similar method can be used to monitor the movement of a cylinder piston rod.

Cylinders are now produced which have grooves in the piston rod covered with chromium. Pulses are generated as the piston rod extends and retracts owing to the magnetic effect varying with the grooves in the piston rod. Two sensors are usually used so that the direction of movement can be established, in the same way as for shaft encoders, and also to increase the resolution of signal, which is typically 0.25 mm (0.010 in).

A cylinder with the sensors fitted inside a hollow rod will give better positional accuracies. This type of cylinder will give pulsed outputs to a resolution of 100 μm (0.004 in).

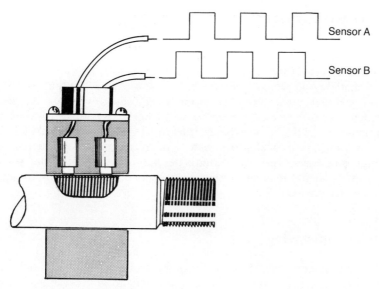

Sensor A

Sensor B

Fig. 5.7   Proximity sensing the outside of a piston rod.

5.4.4 Digital Cameras

A lens is used to focus an object on a tightly packed array of light-sensitive diodes, up to 500 diodes per centimeter length. The object needs to be illuminated, which is best done from behind (Fig. 5.8).

Many applications can be served by using a single row of light-sensitive diodes.

The lens' position can be set to give either magnified or diminished images. A wide range of noncontact measurements are possible at high speed.

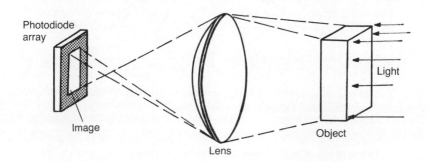

Fig. 5.8   Principle of the digital camera.

5.5. ANALOG DEVICES

An analog device produces a continuous electrical signal with a voltage which varies with external parameters.

Generally, an analog device will give a voltage output which is proportional to its movement. The output is continuous over the full range of movement. Two devices the engineer will come into contact with are the potentiometer and the linear variable differential transformer (LVDT).

To interface an analog device to a microcomputer some signal conditioning is necessary and a converter is needed to change the analog voltage output into a digital reading. This is usually done with an analog to digital converter (see section 6.5).

5.5.1 Potentiometer

The potentiometer (Fig. 5.9) is a variable resistor, and is simple and inexpensive. It can be fitted into a simple circuit to give a variable output voltage as the contact is moved. Points to consider when using potentiometers are:

1.   potentiometers are liable to wear;
2.   if accurate work is to be attempted then potentiometers and their supporting components need to be of high quality.

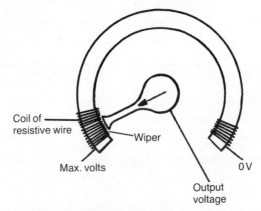

Fig. 5.9   Sketch showing the principle of operation of a potentiometer.

When the choice is available it is better to use digital sensors to interface with the microprocessor because of ease of use.

5.5.2 Linear Variable Differential Transformer

The LVDT works on the principle that the output of a transformer varies as the core is moved. A supporting electronic system is necessary, but this can be an integral part of the LVDT.

The LVDT is constructed (Fig. 5.10a) using two sets of coils, a primary and a secondary coil. The secondary coil has a center tapping or outlet. An alternating voltage is directed into one of the coils while an induced voltage, which is dependent upon the position of the core, is measured in the other coil. The output is taken from the straightest or most linear part of the frequency curve produced by the core displacement (Fig. 5.10b).

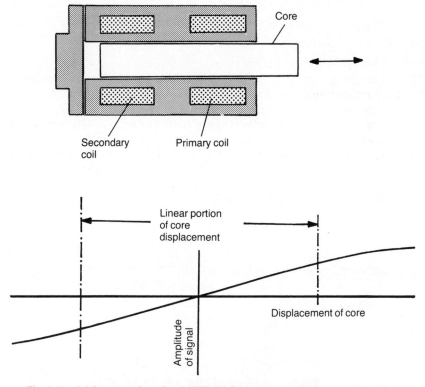

Fig. 5.10   (a) Construction of an LVDT. (b) Frequency to displacement graph.

The features of LVDTs are:

1.   they have little wear;
2.   they come in a vast range of sizes;
3.   they produce a very accurate position signal.

Supporting electronic circuits usually come as an integral part of a LVDT and, like potentiometers, will need an analog to digital converter to interface them with a microcomputer.

There are times when analog devices are the only components which can be used, for example in the case of pressure measurement.

5.5.3   Pressure Transducers

Pressure as defined by force divided by area uses the units of bar or lb/square inch (psi). The bar is 10^5N/m^2 and is almost equal to 14.7 psi (atmospheric pressure); hence its choice as a unit.

Pressure transducers must be selected to fit the range over which they are to work. Fluid power systems can experience great surges of pressure or shock waves, which may be of short duration but are sufficient to destroy the transducer. The ability to withstand these over-pressures must not be overlooked.

Whilst there are various types of pressure transducers (Fig. 5.11) they all have several things in common.

1.   They are all analog devices.
2.   They all have some element, usually a diaphragm, which is displaced or distorted by the application of pressure. This displacement is used to alter some physical property which can be measured. These small variations in physical property have to be converted into representative voltages.
3.   They all need signal conditioning. When ordering pressure transducers it is necessary to determine what signal conditioning elements are necessary. Devices which incorporate these into the body of the tranducers are the easier to use.

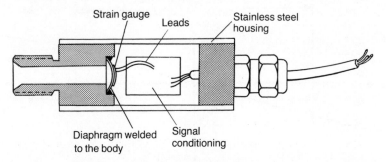

Fig. 5.11   Example of a pressure transducer.

There are three main types of component used for pressure transducers: strain gauges, piezo or quartz crystal and piezo-resistive or integrated circuit type. The features of these three types are described below.

Strain Guage Type

A strain gauge is bonded onto a diaphragm. As the diaphragm is distorted so the resistive value of the strain gauge is altered.

This change in resistance is used to give an indirect measurement of pressure. Pressure transducers based on strain gauges are well established and widely used. They are robust and can withstand transient over-pressures better than other types.

Piezo-electric or Quartz Crystal Type

With pressure changes a piece of quartz crystal produces electronic charges. These charges can be amplified and converted into voltages to represent pressure changes.

Piezo-electric transducers have a very high frequency response and are widely used in applications where rapid changes of pressure are to be measured, such as in the case of shock, vibration and acoustics testing.

Piezo-resistive or Integrated Circuit Type

The principle of operation of the piezo-resistive type is that pressure induces strain in an integrated circuit (IC) which is a part of a silicon diaphragm. This strain is made to generate a voltage which is proportional to the applied pressure.

These devices are more recent in development than the strain gauge or piezo-electric devices, and are widely used for fluid power applications.

They have advantages in cost, size, frequency response and stability. They are unlikely to withstand overloading as well as the strain gauge type.

There are other types of pressure transducers available based upon, say, capacitance or LVDT construction used for measuring the displacement by pressure of a diaphragm. These types are not as widely used for fluid power applications as the types already mentioned.

QUESTIONS

1. What is meant by feedback?

2. What is a limit switch used for?

3. What is an incremental shaft encoder?

4. How can the direction of movement of an incremental shaft encoder be determined?

5. How many positions of rotation will a six track absolute encoder produce?

6. Why is a Gray code used?

7. What is a digital camera?

8. Describe two analog devices used for indicating position.

9. Name the three basic types of pressure transducer used in fluid power applications.

6

Interfacing

Microcomputers usually work with very different levels of voltages and currents from those used by the sensors and actuators they encounter in engineering applications. The means of effecting the connections between items which are not directly compatible is called interfacing. This chapter deals with the techniques most commonly used to interface microcomputers with sensors and actuators found in fluid power applications.

6.1 INTRODUCTION

In order to demonstrate the need for interfacing an exmaple is given in Fig. 6.1. A microcomputer is to control two pneumatic cylinders. Table 6.1 shows typical voltages and currents that are required by the solenoids compared with those available from the microcomputer. Power supplies are needed at the voltage levels of the valve solenoids and interfacing devices. (Table 6.2).

The requirement is for the microcomputer to be able to energize the solenoid with, say, the characteristics in Table 6.1. Patently, the microcomputer is incapable of driving the solenoid directly and an interface is needed. A solution to this problem is given later in this chapter.

Table 6.1   Voltage and current values.

Cylinder	Solenoid requirement	Microcomputer
Out	Energize solenoid with 24 V DC, 0.5 A	Logic level 1,5 V, 50 μA
Stationary	De-energize solenoid 0 V	Logic level 0,0 V

Table 6.2   Various levels of output from the interfacing elements.

Device	Voltage level	Maximum current
Microcomputer port	5 V	0.1 A max.
Relay	6–24 V	1–5 A
Solid state relay	3–5 V	0.003–0.005 A
DC solenoid	12–190 V	0.3–5 A
AC solenoid	100–250 V	450 V A

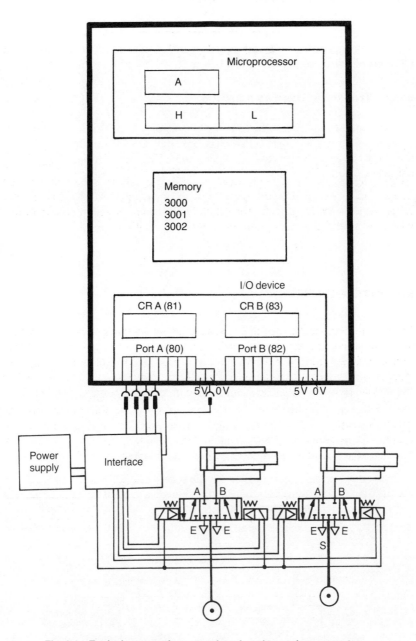

Fig. 6.1 Typical pneumatic system interfaced to a microcomputer.

6.2 COMPONENTS FOR SWITCHING

External devices which require switching, say the solenoid of a valve on and off, can be operated quite straightforwardly. The most commonly used methods of switching are discussed below.

6.2.1 Transistor Used as a Switch

Fundamental to many interfaces is the transistor switch (Figs. 6.2 and 6.3). Used in this way a transistor is an electronic device which switches a current flow between two points C and E. This is done when a small signal is given to a third connection B.

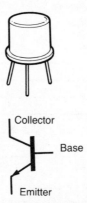

Fig. 6.2 Actual and pictorial representation of a transistor.

The three pins are called emitter, base and collector, and the arrow shows the direction of conventional flow of current.

Used as a switch, the following conditions occur:

1. as an *off* switch, i.e. no voltage at the base;
2. As an *on* switch, i.e. there is a voltage at the base which from a microcomputer would be typically 5 volts.

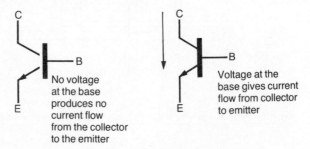

Fig. 6.3 Diagrammatical representations of a transistor.

6.2.2 Transistor Switch and Solenoid

It follows from the above that as a microcomputer can provide a voltage to the base of a transistor, it can switch large currents from an external source. A circuit to switch the solenoid of a relay might be as follows. R is a resistor of, say, 500 ohms which limits the current from the microcomputer, thus saving power and acting as some protection to the port, e.g. switching a solenoid-operated pneumatic valve using a relay (Fig. 6.4).

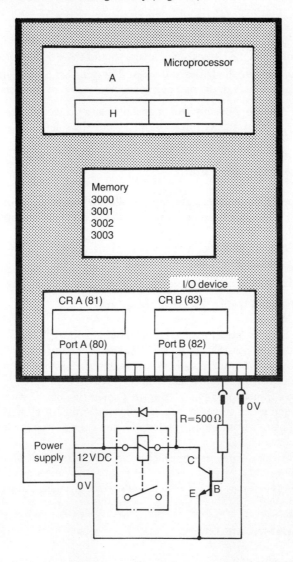

Fig. 6.4 Example of a transistor switching the solenoid of a relay.

Using a relay is a very common method of switching and one widely used arrangement with which the engineer should be familiar. One way of interfacing this with the microcomputer is to use the transistor to switch the solenoid of the relay. Relays exist which will energize from 5 V and will draw 90 mA of current. This would just operate from a microcomputer. A better method would be to use a relay of a larger capacity with, say, a 24 V solenoid and switch this with the transistor.

In Fig. 6.5 the transistor controls a solenoid which is part of a relay. The relay in turn switches a power supply which energizes a solenoid as part of a pneumatic valve. It can be seen that the relay physically separates the large power source (which could be mains electricity) from the microcomputer. This could save the latter from damage should there be a malfunction.

Relays come in all shapes and sizes. They consist of a solenoid which, when energized, closes or opens a pair (or several pairs) of contacts.

NOTE: A diode is usually added as shown as a protection against damage by back 'electromotive force' (e.m.f.) when the solenoid is switched off.

6.2.3 Open Collector Output

Sometimes microcomputers have 'open collector' outputs; in other words, they have the transistor built in. If the port of the previous example had been an 'open collector' it would be connected as shown in Fig. 6.6.

The user must provide the power supply to energize whatever device is being used, typically relays, solenoids or electric motors.

6.2.4 Switching Using a Power Transistor

The typical current drawn by a valve solenoid is approximately 3 A. A small transistor cannot possibly pass such current without destroying itself. However, just as the small transistor can switch the solenoid of a relay, so it can switch a transistor of much larger size.

A power transistor is made up from a small transistor switching a large transistor; such a device could be a 'Darlington' transistor. This is a device containing a pair of transistors. Figure 6.7 shows the previous example using this means of switching.

Each of these three examples (Figs. 6.5, 6.6 and 6.7) show the same valve being switched to create a flow of fluid. Only a system using a relay would be suitable for a solenoid which is energized by alternating current.

6.2.5 Thyristors and Triacs

For switching alternating current, thyristors and triacs are used. A thyristor or silicon controlled rectifier (SCR) is like a diode with a gate control. It will act like a diode and continue conducting current after the gate has received a voltage pulse. If the current ceases, even momentarily, it will no longer conduct until the gate receives another voltage pulse.

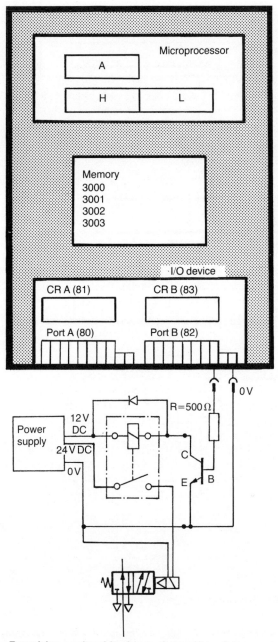

Fig. 6.5 Energizing a solenoid using a relay and transistor as a switch.

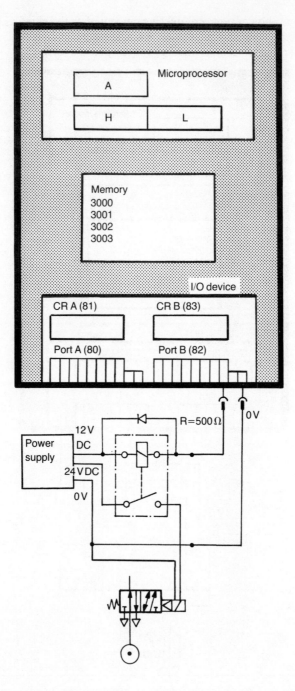

Fig. 6.6 Energizing a solenoid using a relay connected to the microcomputer port.

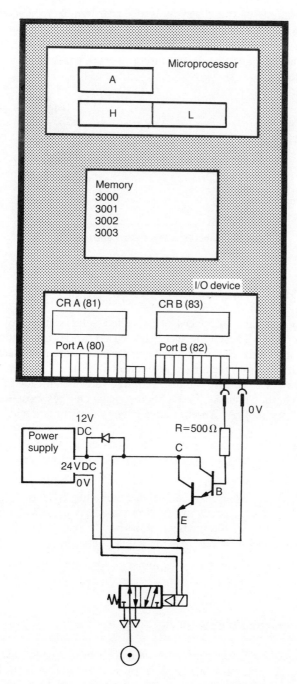

Fig. 6.7   Energizing a power transistor to operate the solenoid of a valve.

A development of the thyristor is the triac which behaves like two thyristors connected together.

Thyristors and triacs (Fig. 6.8) are mainly used for switching and regulating industrial plant and electric motor control, particularly alternating current applications.

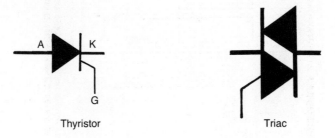

Thyristor   Triac

Fig. 6.8   Symbols of thyrsistors and triacs.

6.2.6   Solid State Relay Interfaces

A family of proprietary interfaces often called solid state relays is now available (Fig. 6.9). These interfaces are based upon transistors, opto-isolators, thyristors and triacs. Physically, these solid state relays are usually small blocks with just four terminals (see section 6.2.8).

It is necessary to specify the load required (e.g. magnitude of solenoid current) and whether it is AC or DC before use. Devices exist which will switch currents up to 40 A. Another advantage is that these devices are also opto-isolated.

Generally, they are an elegant way of interfacing. They are sometimes rejected for the following reasons compared with conventional relays:

1.   They are more expensive.
2.   They are silent. Only a relay can be heard clicking as it works.
3.   They will fail completely when overloaded. A relay's contacts will become pitted but will probably still work.
4.   They cannot operate several pairs of contacts at once.

However, since they have no moving parts they should function indefinitely.

6.2.7   Light-emitting Diode

Light-emitting diodes (LEDs) are used extensively in microcomputer control circuits to indicate whether a signal is present or not (Fig. 6.10). For example, if a cylinder piston fails to move when expected, an LED in the circuit would give some indication where the fault may lie.

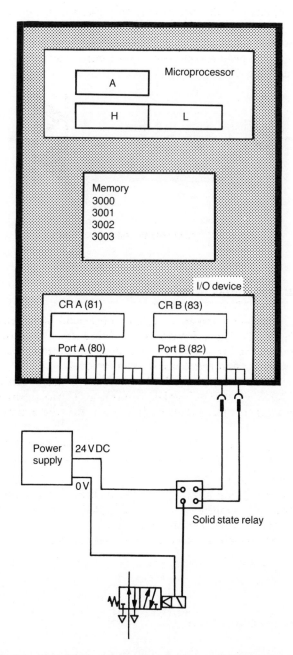

Fig. 6.9   Energizing a valve solenoid using a solid state relay.

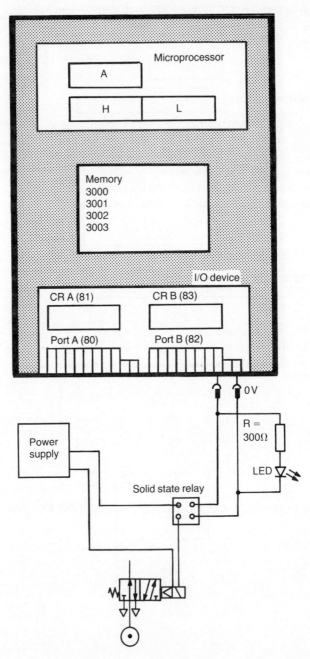

Fig. 6.10   Energizing a valve solenoid using a solid state relay with an LED as an indicator.

LEDs give out light using very little power and micrcomputers can drive them easily.

A 5 V supply needs a resistor of about 300 Ω to limit the current flow, to prevent the LED being destroyed.

6.2.8 Opto-isolators

Signals are transmitted by driving a LED which in turn switches a light-activated switch or phototransistor. They may act as an interface in themselves in order to make compatible different voltage levels. They also serve to isolate elements from electronic noise transmission. They can be used for both *input* and *output* signals to the microcomputer. They are recommended for isolating microcomputers from external disturbances.

6.3 INTERFACING MECHANICAL SWITCHES

Mechanical switches such as push buttons, pressure switches and limit switches can be connected directly to the microcomputer in the manner shown. The power supply of the microcomputer is used to give terminals of 0 and +5 V. It is usual to leave an input always high at +5 V and to drop to 0 V when a device is operated. To limit the current drawn from the microcomputer, a resistor of approximately 5 kΩ is inserted between the positive supply and the 'bit' of the port (Fig. 6.11).

Closing the contacts of the switch connects the 'bit' of the port to 0 V. The resistor limits the current drawn to approximately 1 mA.

6.4 CONVERTING DIGITAL VALUES TO ANALOG VOLTAGES

All information exchanged with the microcomputer has to be digital. This means the data leave or enter the microcomputer via a number of wires, each of which can be in one of two states, known as either logic 0 (0–0.8 V) or logic 1 (2–5 V). Devices for use with the microcomputers fall into two categories.

DIGITAL: the device can only be in one of a set number of states, e.g. a solenoid is either switched on or off, i.e. energized or de-energized.

ANALOG: the device can be in any continuous condition between limits, e.g. a proportional hydraulic valve will give a variable flow from a continuous range of input values.

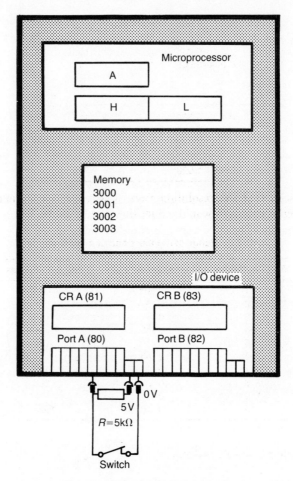

Fig. 6.11   **Simple switch interfaced with a microcomputer.**

6.4.1   Digital Considerations

Microcomputers are themselves digital devices and surprisingly many of the devices they interface with are digital. Numbers are represented in binary form by the states of the wires.

Table 6.3 gives the number range which can be represented by different numbers of wires.

EXAMPLE: a linear position device having a range of 1 cm is presented to a microcomputer through 8 wires or 256 different conditions. The resolution of movement is shown for the representative numbers.

Displacement centimeter	Number presented
0−0.004	0
0.004−0.008	1
.	.
.	.
.	.
0.496−0.500	127
.	.
.	.
.	.
0.996−1.000	255

It is seen that the resolution from 8 wires is 0.4%. Increasing the number of wires to 9 would reduce the resolution to 0.1%

Table 6.3   Range of numbers and number of states.

Number of wires	Range of numbers	Number of states
1	0 and 1	2
2	0−3	4
3	0−7	8
4	0−15	16
5	0−31	32
6	0−63	64
7	0−127	128
8	0−255	256
9	0−511	512
10	0−1023	1024

6.4.2   Analog Considerations

The various devices for measuring pressure, flow and temperature are usually analog. To exchange this information with the computer the analog signal has to be digitized so that it is compatible with the microcomputer. Devices to interface these components with the computer are described in section 6.5.

6.4.3   Advantages and Disadvantages of Digital over Analog

The advantages of digital over analog are that:

1.   the values cannot drift, i.e. appear to change by themselves;
2.   the interfacing is simpler;
3.   they are less susceptible to interference.

The disadvantages are;

1.   the resolution can be limited;
2.   absolute values may be difficult to obtain.

6.5   ANALOG TO DIGITAL AND DIGITAL TO ANALOG CONVERTERS

Analog to digital and digital to analog converters convert voltages from analog values to digital numbers, A/D, or vice versa, D/A. The resolution of any system depends upon the number of bits used to describe the value. The accuracy of that value depends upon the internal/external voltage reference. Standard digital to analog converters are used to satisfy a voltage and not a current. If current is required for, say, a proportional valve, then an amplifier is needed to change both the voltage levels and the power supplied (Fig. 6.12).

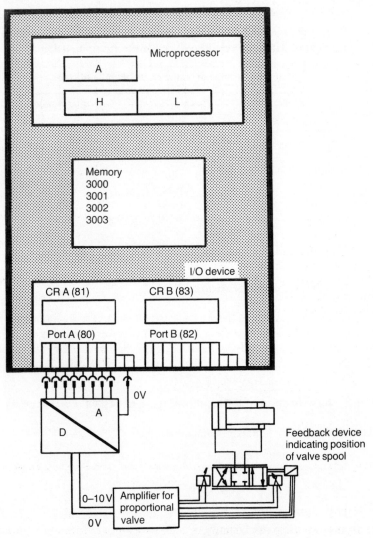

Fig. 6.12   Analog ouput to a proportional valve.

A/D converters need the input to be within a fixed range. An 8-bit A/D typically has the following characteristics (Fig. 6.13):

1.   0 V gives number 0 as output;
2.   2.5 V gives number 255 as ouput.

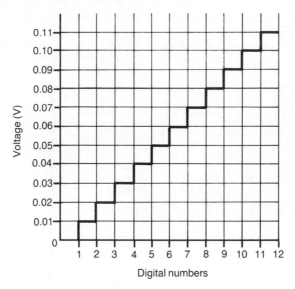

Digital numbers

Fig. 6.13   Graph showing part of the stepped output to produce 0–2.5 V.

If a different range of input voltages is required, some signal conditioning or amplifying must be done.

To produce a digitized input from an analog signal, say from a temperature sensor, the example above would be reversed (Fig. 6.14). The signal would be conditioned.

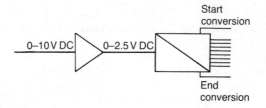

Fig. 6.14   Electronic circuit showing the principle of converting analog signals to digital signals.

The engineer does not need to be concerned about the detail of the electronics as most manufacturers of microcomputers produce interfacing boards which integrate with their microcomputer systems.

6.6 PULSE-TIME TO ANALOG CONVERSION

Another method of producing an analog input to or output from a microcomputer is to use a pulse-time to analog converter (Fig. 6.15). This is a form of A/D and D/A converter, since the timing concerned is the time of a pulse in a pulse train. Microprocessors can easily both generate pulse trains and monitor them.

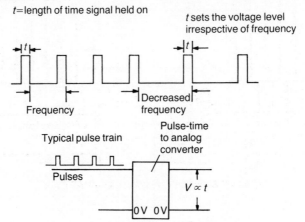

Fig. 6.15 Analog output from a pulse train.

6.7 SERIAL TRANSMISSION

If the microcomputer is some distance from the system under control, it is patently undesirable to have lots of parallel wires to transmit data. A solution is to use universal asynchronous receiver transmitters (UARTs).

Figure 6.16 shows information coming from the microcomputer, but a similar configuration can transmit data to the microcomputer. Parallel information goes into the UART register. This information is then transmitted as a series of pulses to the receiving UART, where it is again available as parallel data. As far as a user is aware, the microcomputer is connected directly to the fluid power system. There are small time lags which may have to be allowed for.

UARTs have timing devices to coordinate the transmitting and receiving of data which need to operate at roughly the same rate (baud rate).

Some microcomputers have UARTs as part of the input/output facilities. The serial connection between the UARTs can be by a screened cable or a 'twisted pair', if interference is not too important. A 'twisted pair' is simply a couple of wires twisted around one another.

For very long distances, or where interference is a problem, or the environment toxic, a fiber optics link may be necessary (see section 7.3).

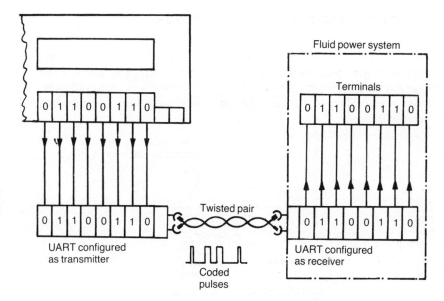

Fig. 6.16 Serial transmission using UART's.

QUESTIONS

1. Explain why interfacing is necessary.

2. A microcomputer is to be used to switch on and off a solenoid operated valve. Sketch an interface between the microcomputer and solenoid using a transistor and a relay.

3. What are LEDs?

4. How can LEDs be used in fluid power applications?

5. How are LEDs inserted into a circuit?

6. Sketch a circuit showing a limit switch interfaced to a microcomputer.

7. What are opto-isolators and why are they used?

8. What are solid state relays?

9. Use the 8 bits of a microcomputer port to represent a voltage range of 0–12 V. Confirm that the binary code 10011011 represents 7.3 V. Confirm that the increments in voltage would be 0.047 V.

7

Interference and Noise Suppression

Interference or electrical noise can affect the working of a microcomputer in several ways. The operation of the microcomputer itself can be disturbed. Interference can give extra pulses which result in specious information being considered. The very small analog signals from a transducer can be easily distorted by interference.

7.1 THE PROBLEM

Interference or electrical noise has much the same characteristics as acoustical noise, in that it is intermittent and its source is difficult to locate. The techniques for curing it range from the simple to the highly complex. It requires skills more appropriate to an art than a science.

A fairly unsophisticated control program might contain 500 bytes, i.e. 4000 bits. Corruption of only one bit could be sufficient to make that program inoperative. Control could be lost by distortion of one of the many internal signals guiding the microprocessor through the program.

Many early microcomputer systems were found to be unsuitable because of interference. Now systems are much more resilient and also robust in their construction, both from the point of view of the electronic components and the made-up boards themselves. They can be used in many situations were interference is prevalent. Even so, a number of possible steps should be taken to prevent any likely interference having adverse effects.

In any installation, interference is likely to be present. The causes are numerous, typically from electromechanical equipment such as electric motors, electric-welding equipment, relays and switches. Interference will be radiated to the microcomputer or conducted through any wires. Wires themselves act as aerials and interference can be induced in them. The main power supply can be the source of interference with the cause of contamination many miles away; for example, a flash of lightning can induce voltage surges in overhead power cables.

7.2 STANDARD PRECAUTIONS

In any situation interference is likely and the microcomputer should have the following as standard.

1. The system should be correctly earthed and screened, i.e. surrounded by an earthed metal plate or enclosed in a metal cabinet.
2. The components inside the system should be correctly decoupled, that is filtered by capacitors, especially the microprocessor.
3. The mains must be filtered. A filter is normally incorporated as part of the microcomputer DC power supply where it is fed by the AC mains. Proprietary filters for this task are inexpensive and readily available.
4. All inputs and outputs to the microcomputer must be opto-isolated. (See section 6.2.8).
5. Programs should be in some type of ROM, not RAM or disc. (See section 4.5.).
6. All leads should be kept as short as possible as they can act as aerials which will pick up interference. Screened leads will help to eliminate this, but they must be connected at one end only to Earth.
7. Ground loops must be avoided for instrumentation as the potential of the Earth varies from point to point owing to the changes in the mains electricity supply. If a system is earthed at several points then there is no common voltage datum and so a ground loop occurs.

Good proprietary systems incorporate all of these features in their construction.

7.3 EXTRA PRECAUTIONS

If interference is still a problem or likely to be, then further actions must be taken.

7.3.1 High Quality Filter

It may be necessary to have a high quality filter near the mains plug as well as the standard built-in filter. The design of filters is a specialized subject and proprietary devices are recommended.

7.3.2 Capacitors

A cheap, simple and effective way of eliminating interference can be by the use of capacitors. In simple terms, capacitors offer little impedance (rather like resistance) to alternating current; the higher the frequency the less the impedance. Capacitors have an infinite impedance to direct current. Noise may be considered to be high frequency.

EXAMPLE: A binary signal from a light-activated switch is contaminated by noise. A capacitor across the line removes the high frequency noise content. (Fig. 7.1).

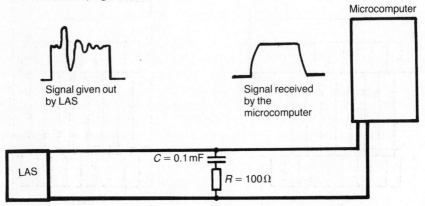

Fig. 7.1   The effect that a capacitor has in absorbing high frequency noise.

NOTE: The capacitor not only removes the noise but it also rounds off the edges of the square waveform as the edges represent a high frequency content.

Usually there has to be a compromise, and values of the capacitor are arrived at by trial and error, possibly looking at the resultant signal on a cathode ray oscilloscope. When the interference is of the same order as the clear signal things are more difficult.

7.3.3   Capacitor plus Resistance

In situations where there is arcing, such as in heavy duty relays, capacitors are used with resistors to absorb the voltage pulse.

7.3.4   Fiber Optics

In really difficult situations the microcomputer control element would have to be remote from the system under control and a special link, using fiber optics, would have to be fitted between the two (Fig. 7.2). Light is transmitted down a flexible, narrow, glass tube by total internal reflection and distances of about 26 km can be achieved. Using plastic reduces this distance to about 8 km but the cost is much reduced and plastic is usually more than adequate for fluid power applications. For reasons of both cost and elegance, data are transmitted serially and UARTs are used. (See section

6.7.) Fiber optics are very robust and are resistant to corrosion but they snap if bent sharply.

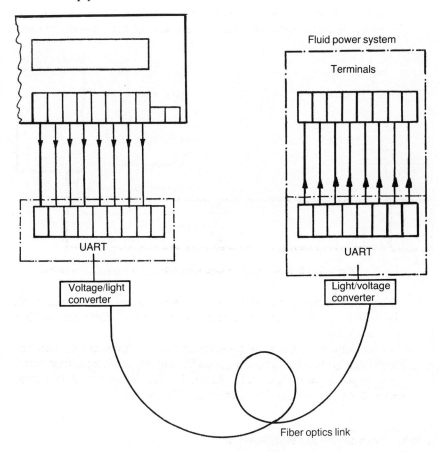

Fig. 7.2   A typical layout of a fiber optics link between the microcomputer and the fluid power system.

7.4   SOFTWARE SOLUTIONS

The procedure is to check a piece of information several times to ensure it is definite. In really extreme cases, say where life is at risk, it is necessary to have several microcomputers to verify one another.

The best known and most widely used example is that of allowing for switch bounce, and normally this should be done for all mechanical switches.

When a switch is opened or closed it does not do so cleanly and it appears to the microcomputer that the switch has been operated more than once. The contacts can bounce for as long as 20 ms. This can be allowed for by the sort of software routine given in Fig. 7.3, where the microcomputer is waiting for the switch to be operated before coming out of a program loop. It checks that the switch is firmly operated and that it has not received a spurious piece of information.

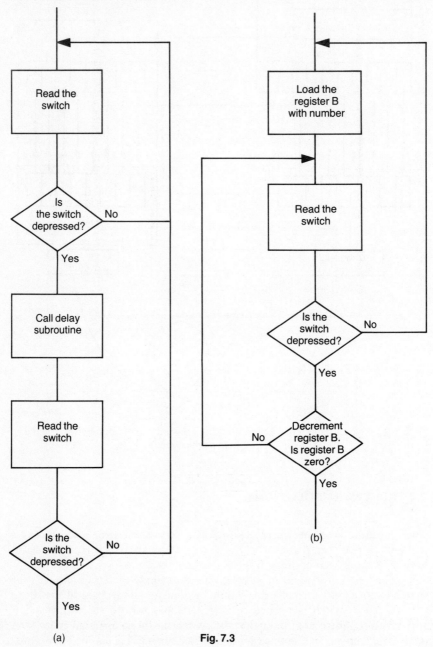

Fig. 7.3

The delay is necessary as microcomputers are getting increasingly faster in their operation. A spurious pulse of a few millionths of a second could be read as positive information.

An alternative is to test the switch many times and only accept the information when, say, 100 consecutive readings are the same (see Fig. 7.3b).

QUESTIONS

1. List causes of interference to microcomputers.

2. List means of eliminating or minimizing interference to a micrcomputer.

8

Programming

This chapter discusses two types of programming language, machine code and BASIC. It is not intended to provide a comprehensive text on programming as many books exist which deal exclusively with this subject. However, there is sufficient information to enable machine code to be used effectively for many fluid power applications.

It is assumed that BASIC is understood by the reader as this is taught in many schools and it is widely used by home computers.

8.1 TYPES OF PROGRAMMING

A program is a sequence of instructions which the microcomputer is given to execute. The choice is between a high and a low level language or a mixture of the two.

HIGH LEVEL: BASIC is the language most widely used on microcomputers but many others exist.

LOW LEVEL: machine code is the code the microprocessor executes. Assembly language is a symbolic or mnemonic form of machine code. Both machine code and assembly language are commonly referred to as 'machine code'.

For examples of instructions for these languages see sections 8.6.1 and 8.8.1

Assembly language requires translation into machine code before use. It can be done manually by the programmer or by means of a program called an assembler.

A high level language must also be translated into machine code. This is done by a program called a 'compiler', or alternatively executed at the time of use by a program called an 'interpreter'. The interpreter must be present in the microcomputer and is usually stored in PROM.

8.2 COMPARISONS BETWEEN MACHINE CODE AND BASIC

As a general rule, programming is most easily and cheaply done in as high a level language as possible. Thus BASIC is easier to use than machine code. A program which is written in as high a level language as possible is the easiest, and therefore less costly, to produce.

In applications concerning control and monitoring of machinery, it is invariably found that certain portions of the program require precise timing or extreme speed. In these circumstances, it may be necessary to write these portions as subroutines in machine code and then insert them into the program written in the high level language. As processing speeds of microprocessors increase so the proportion of programming needed to be done in machine code will decrease.

A comparison between the two levels of programming languages is given in Table. 8.1. Figures given in this table are typical values based on the authors' experiences and will vary widely from microprocesor to microprocessor. It can be seen that there are distinct advantages with each language and hence there is a need to be able to use both.

Table 8.1   Comparison of machine code and BASIC.

Machine code	BASIC
Fast, each instruction 0.5 μs	Slow, each instruction 1 ms
Timing in increments of 0.1 μs to quartz crystal accuracy	Timing unknown
Difficult to learn and use, approx. 40 days	Easy to learn and use, approx. 2 days
Programs slow to develop, say 20 hours	Programs fast to develop, say 30 minutes

8.3 REGISTERS

Before a microprocessor can be programmed it is necessary to understand a little about how it works.

Integrated circuits consist of electronic circuits etched on to tiny pieces of silicon. Microprocessors are described as large scale integrated circuits. This is due to the many thousands of tiny electronic circuits from which they are made. However, to use them they are represented as groups of registers each with its own purpose.

Typically, a register (Fig. 8.1) consists of eight bits (analogous to eight pigeon holes). They are numbered from 7 to 0. Each bit can be at a voltage level which is either high (approximately 5 volts) considered as logic 1, or low (approximately 0 volts) and considered as logic 0 (Fig. 8.2).

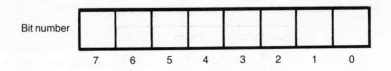

Bit number

7   6   5   4   3   2   1   0

Fig. 8.1   Representation of a register.

The binary number 10011110 shown in Fig. 8.2 is very cumbersome to use. It could be converted into the familiar decimal number 158 but it is much easier and more efficient to use *hexadecimal* numbering.

Hexadecimal is used for most programming applications, but occasionally when considering individual bits, binary digits are used.

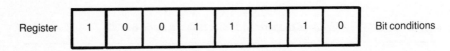

Register   1   0   0   1   1   1   1   0   Bit conditions

Fig. 8.2   Logic levels of a typical register.

8.4   HEXADECIMAL NUMBERING SYSTEM

As the hexadecimal number system is to a base of 16 it runs out of digits after counting to 9. To complete the count up to 16 the letters A, B, C, D, E, F are used. Just as in denary after 9 the number is one–zero or 10, so in base 16 after the letter F the number is one–zero. When reading hexadecimal numbers (HEX), 10 should always be understood as one–zero and not ten, 20 as two–zero, etc.

8.4.1   Positive Numbers

Comparison of hexadecimal (base 16) and denary (base 10) and binary numbering is shown in Table 8.2.

Table 8.2 Representation of positive numbers.

Hexadecimal	Denary	Binary
1	1	0001
2	2	0010
3	3	0011
4	4	0100
5	5	0101
6	6	0110
7	7	0111
8	8	1000
9	9	1001
A	10	1010
B	11	1011
C	12	1100
D	13	1101
E	14	1110
F	15	1111
10	16	10000

8.4.2 Negative Numbers

Negative numbers are shown in Table 8.3. Consider the analogy of a mileometer being turned back too far: the first overshoot shows up as 999 (denary) which is -1, -2 is 998, -3 is 997 and so forth.

In HEX, -1 is FF, -2 is FE and -3 is FD. FF can therefore represent 255 (denary) or -1 depending on the circumstances.

Table 8.3 Representation of negative numbers.

Hexadecimal	Denary	Binary
FF	-1	11111111
FE	-2	11111110
FD	-3	11111101
FC	-4	11111100
FB	-5	11111011
FA	-6	11111010
F9	-7	11111001
F8	-8	11111000
F7	-9	11110111
F6	-10	11110110
F5	-11	11110101
F4	-12	11110100
F3	-13	11110011
F2	-14	11110010
F1	-15	11110001
F0	-16	11110000
EF	-17	11101111

8.4.3   Converting Binary Numbers into Hexadecimal

There are various methods of acquiring the hexadecimal numbers; one such is the 'remainder method'. One favored by the authors is to place the numbers 8, 4, 2, 1 above each set of four digits. Then by adding the numbers adjacent to the 1 in the binary format the hexadecimal number is found.

EXAMPLE: To convert 11000111 into HEX format, write the numbers 84218421 in two groups of four above the binary number as shown:

$$8\ 4\ 2\ 1 \qquad 8\ 4\ 2\ 1$$
$$1\ 1\ 0\ 0 \qquad 0\ 1\ 1\ 1$$

By adding the numbers adjacent to 1 together we get:

$$\text{Denary} \quad \text{HEX}$$
$$8+4=12 = C$$
$$4+2+1 = 7$$

$$\text{Answer} = C7$$

ANOTHER EXAMPLE: To convert 11010110 into hexadecimal numbers:

$$\text{As before} \quad 8\ 4\ 2\ 1 \qquad 8\ 4\ 2\ 1$$
$$1\ 1\ 0\ 1 \qquad 0\ 1\ 1\ 0$$

$$\text{Denary} \quad \text{HEX}$$
$$\text{Then} \quad 8+4+1=13 = D$$
$$4+2 = 6$$

$$\text{Answer} = D6$$

8.5   CHOICE OF MICROPROCESSOR

Before any programming is done it is necessary to know what equipment one is dealing with. For control purposes the two most important elements are the microprocessor and the input/output device.

8.5.1   Microprocessor

In this book the Z80 microprocessor, manufactured by ZILOG, has been chosen on which to base the worked solutions. The Z80 is a widely used microprocessor for control applications and is likely to remain so.

The principles given for arriving at solutions are applicable to whatever microprocessor might be used.

Two registers of the Z80 microprocessor are shown in the following example; others, like B and C, exist (Fig. 8.3). These are just part of the architecture which can be compared with the full set of registers which is shown in appendix C.

EXAMPLE:

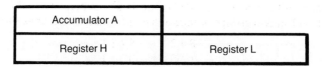

Fig. 8.3   Internal registers of a microprocessor.

Accumulator: This is the register through which most data passes. It holds eight bits which can of course be represented by two HEX numbers.

HL (high/low): This is a general purpose register. It can be used as one register HL, or independent as two 8-bit registers H and L.

8.5.2   The Input/Output Device

In order to connect to external components an input/output device is required. One that is compatible with the Z80 microprocessor is the PIO (parallel input/output). A simplified layout is shown in Fig. 8.4.

There are two ports, A and B, of eight bits each which can be used for either taking signals in or giving signals out. The mode of operation of each bit is determined by the contents of the appropriate control register, CR.

The hexadecimal numbers 80 to 83 are typical addresses used for the ports A and B and the control registers.

8.6   MACHINE CODE PROGRAMMING

In order to program a microcomputer in machine code it is necessary to understand the instruction set. This is the list of commands that the microprocessor is capable of performing. Initially it seems to be a daunting task to understand. However, once the means of developing even the simplest of programs is understood, subsequent programming is much less difficult.

Fortunately, control applications require only a small selection of commands from the instruction set, say 30 out of the possible 1000. A selection of these are given here; for details of the full set see appendix C.

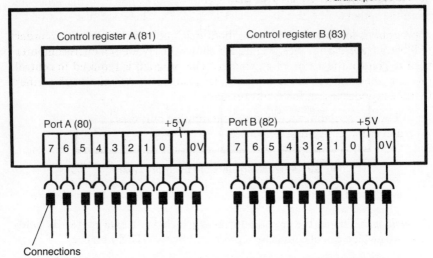

Fig. 8.4   Simplified layout of a parallel input/output device.

8.6.1   Limited Machine Code Instruction Set

The meaning of these instructions will become clear by working through an example.

Machine code	Assembly language	Explanation of language code
3E _ _	LDA,n	Load accumulator A with data n.
3A _ _ _ _	LDA,(nn)	Load A with data at address nn.
32 _ _ _ _	LD(nn),A	Load address nn with data in A.
21 _ _ _ _	LD HL,nn	Load register HL with data nn.
2B	DEC HL	Decrement contents of HL.
BC	CP H	Compare data in H with data in A.
B7	OR A	OR A with itself.
C3 _ _ _ _	JP nn	Jump to address nn.
18 _ _	JR,DIS	Jump relative.
28 _ _	JRZ,DIS	Jump relative if result zero.
20 _ _	JRNZ,DIS	Jump relative if result not zero.
06 _ _	LDB,n	Load register B with data n.
10 _ _	DJNZ,DIS	Decrement register B. Jump relative if result not zero.
D3 _ _	OUT(n),A	Load I/O port at address n with A.
DB _ _	IN A,(n)	Load A from I/O port at address n.
CD _ _ _ _	CALL nn	Call subroutine at address nn.
C9	RET	Return from subroutine.

8.7 PROGRAMMING EXAMPLE

A program is to be developed using both machine code and BASIC in order to illustrate these languages. Figure 8.5 shows the layout of a microcomputer used to control the action of a cylinder. The program is required to control the extending and retracting of a cylinder piston rod. A limit switch verifies that the cylinder piston rod is fully extended.

The essential information for programming purposes is:
1. a microprocessor based upon Z80;
2. some RAM memory with starting address at 3000 in which to develop a program;
3. an input/output based upon PIO as described above—port A is to be designated as output and port B as input (the method of doing this is described in appendix D);
4. interfaces between the microcomputer and the valves (for further information see chapter 6);
5. valves to select the direction of movement, in this example pneumatically operated;
6. cylinder;
7. limit switch.

8.7.1 The Stages of Machine Code Program Development

There are usually four stages of program development:
1. define the sequence of operation;
2. draw a flow chart;
3. write the assembly program;
4. convert to machine code.

1. Define the Sequence of Operation
It is not always easy to decide exactly what the microcomputer has to do. For the sake of simplicity an elementary example has been chosen. The requirement is to program the microcomputer to do the following:
(a) Provide a signal of +5 V from bit 1 of port A to send the piston rod of a cylinder out.
(b) Determine when the cylinder piston rod is extended by reading the signal from the limit switch at bit 0 of port B.
(c) Clear the signal from port A to de-energize the solenoid and retract the cylinder.

2. Draw the Flow Chart
The sequence of operations is now converted into a flow chart (Fig. 8.6). The symbols used are standard, with a rectangle signifying a general statement, and a diamond signifying a branch or conditional jump.

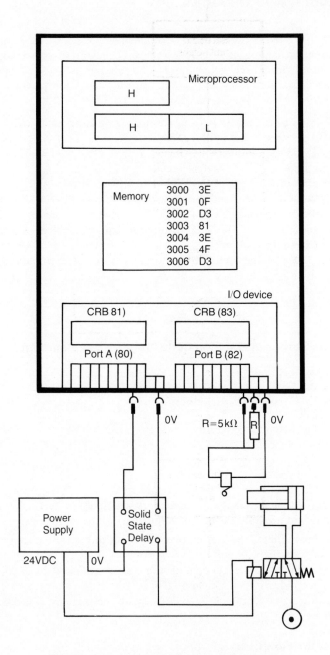

Fig. 8.5 Typical microcomputer control system.

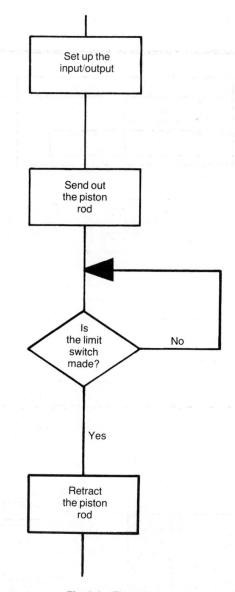

Fig. 8.6   Flow chart.

Further description:

SET UP THE I/O: the I/O device must be configured to define the inputs and outputs as required. This is done using a set of rules laid down by the manufacturer of the microcomputer. The addresses for the control registers and the ports are also laid down by the manufacturer.

In this example the device is based upon the PIO (see appendix D) and requires setting up as follows.

(a) CR for port A is loaded with OF Hex.
(b) CR for port B is first loaded with CF Hex, then with FF Hex.

SEND OUT THE PISTON ROD: port A has to be loaded with a number to make bit 1=1, i.e. +5 volts.

IS THE LIMIT SWITCH MADE? This is a branching operation. For the Z80 microprocessor a logic instruction OR A is necessary.

RETRACT THE PISTON: port A has to be loaded with a number to make bit 1=0, i.e. 0 volts.

3.   Write the Assembly Program
This is done using the instruction set provided by the microprocessor manufacturer. Lots of comments after the semicolon should be included to explain what is happening and as a reminder to the programmer.

Assembly language	Comments
LDA,OF	; Load accumulator A with OF.
OUT(81),A	; Load CR A at address 81 with contents of A.
LDA,CF	; Load accumulator A with CF.
OUT(83),A	; Load CR B at address 83 with contents of A.
LDA,FF	; Load accumulator A with FF, all inputs.
OUT(83),A	; Load CR B at address 83 with contents of A.
LDA,02	; Load A with 02(00000010 binary).
OUT(80),A	; Load port A address 80 with contents of A, this makes bit 1 of port A high or at +5V, and energizes the solenoid.
IN A,(82)	; Port B is loaded into the accumulator A.
OR A	; Logic statement.
JR Z, DIS	; Jump relative if result is not zero, this operation reads bit 0 of port B until the limit switch makes the value zero or 0V. The displacement is FB (-5)
LDA,00	; Load A with 00 (00000000 binary).
OUT(80),A	; Load port A at address 80 with content of A. This switches the solenoid off and retracts the piston rod.
HALT	

4.   Convert into Machine Code
The assembly language must be converted into machine code (hexa-decimal notation) and addresses of memory assigned. The hexadecimal notation for the machine code conversion from assembly language is desig-nated by the manufacturer of the microprocesor (see appendix C and section 8.6.1.)

Address of memory	Machine code	Assembly language
3000	3E	LDA,OF
3001	0F	
3002	D3	OUT(81),A
3003	81	
3004	3E	LDA,CF
3005	CF	
3006	D3	OUT(83),A
3007	83	
3008	3E	LDA,FF
3009	FF	
300A	D3	OUT(83),A
300B	83	
300C	3E	LDA,02
300D	02	
300E	D3	OUT(80),A
300F	80	
3010	DB	IN A,(82)
3011	82	
3012	B7	OR A
3013	20	JRNZ,FB
3014	FB	
3015	3E	LDA,00
3016	00	
3017	D3	OUT(80),A
3018	80	
3019	76	HALT

8.8 PROGRAMMING IN BASIC

There are many types or dialects of the BASIC programming language used for microcomputers, each with its own specific instructions.

Most users of BASIC know the instructions INPUT, LET, PRINT, etc. but the more obscure like POKE, OUT, NOT, OR, AND, INP, are less familiar. For this purpose, Table 8.4 explains some of these instructions.

8.8.1 Some BASIC Commands and Instructions

Table 8.4 will be helpful for writing programs in a high level language.

These BASIC commands and functions are correct for the CP/M version of Microsoft BASIC; other versions of BASIC could differ slightly.

Table 8.4 BASIC commands.

Command	Meaning
POKE	Store the specified byte at indicated memory address.
OUT	Put the specified byte to the output port.
NOT	Logical function which inverts the logic value of a byte.
INP	Read the value of the port.
OR	Logic function. See appendix F.
AND	Logic function. See appendix F.
PEEK	Read the contents of the memory location which follows.
CALL	Calls a subroutine.
&H	Serves as prefix for hexadecimal constant.

8.8.2   The Program in BASIC

The previous example of a control system is now written in BASIC and illustrates the difference between machine code and BASIC.

```
10   OUT &H81,&H0 F ;SET UP I/0   (&H=HEXADECIMAL)
20   OUT &H83,&HCF
30   OUT &H83,&HFF
40   OUT &H80,2       ;SEND CYLINDER PISTON ROD OUT.
50    IF IN(&H82)=0   THEN GOTO 50; WAIT FOR THE LIMIT
                             SWITCH TO BE CONTACTED.
60   OUT &H80,0       ;SEND CYLINDER PISTON IN.
70   END
```

It is easy to appreciate how much simpler BASIC is to write than assembly language and machine code.

8.9   POINTS TO NOTE ON THE ABOVE PROGRAMS

1.   There is no particular speed requirement in this example and therefore BASIC serves perfectly well.
2   For simplicity the ports have been dealt with using all the eight bits at once.
3.   If there had been several cylinders and limit or start switches, then the concern would have been with individual bits.
4.   For further information relating to bit handling see appendixes E and F.
5.   Further programming examples are given in chapters 11 and 12.

QUESTIONS

1. Give examples of instructions in machine code, assembly language and BASIC.

2. Convert the binary number 11000101 into hexadecimal.

3. Convert the hexadecimal number C4 into a binary number.

4. Develop a piece of machine code program which would put the binary number 10100111 at an output port with address 4A.

9

Proportional Solenoid Valves

One major development in fluid power has been the production of a low cost, electrically modulated valves. This has become known as the proportional solenoid valve. For many years acceleration and deceleration of loads were achieved by using a combination of pressure and flow control valves. Now it is possible to control both of these operations by using a proportional solenoid valve. Proportional solenoid valves are more practical for the control of load movement and they also give cost savings compared with using other standard types of valve. In this respect developments in pneumatic valves have not kept pace with those for hydraulic valves.

9.1 INTRODUCTION

Proportional solenoid valves are similar in construction to pressure, flow, and directional control valves. Proportional pressure control valves vary the load pressure by measuring the electrical current given to the solenoid, whereas normal pressure control valves vary the pressure by a mechanical adjustment.

There are two main types of proportional solenoid valves; pilot-operated valves and direct-operated valves. Proportional solenoid valves are used for the following applications:

1.   pressure control;
2.   flow control;
3.   directional control.

Proportional solenoid valves are more accurate than normal valves which are used to set the pressure, flow, and direction of fluid movement.

The solenoids used for proportional valves tend to be constructed in different ways. One type of solenoid is current-controlled and has a short stroke. The other type is voltage-controlled and has a long stroke. Describing the solenoids by either voltage or current is misleading as both types need current to produce the output force; therefore, using the stroke length is a better way to distinguish them.

The accuracy of proportional solenoid valves is derived from the electronic circuits which control them. The accuracy of a typical amplifier card is 5 mA/V of input, which will give an accuracy of approximately 50 mA with 10 V input signal. With short stroke proportional valves, which operate from a maximum current of 700 mA, the accuracy between input to output of the valve will be of the order of 7%.

Positional control of pneumatic systems is difficult because of the high compressibility of air. They are limited to slow response applications.

9.2 SHORT STROKE SOLENOID VALVES

Short stroke proportional solenoid valves tend to be used as pilot valves because they usually have strokes of only 1–1.5 mm (0.06 in). This restriction in stroke length limits the amount of fluid which will pass through the valves.

Consideration is now given to the different types of proportional solenoid valve.

9.2.1 Pressure Control Valves

Pressure control valves are designed to give adjustment to the poppet which limits the pressure of the system. This is done by a solenoid, as opposed to a manual adjustment.

Figure 9.1 shows a solenoid controlling a pilot valve which will operate the main stage of a pressure control valve. Compared with a normal pressure

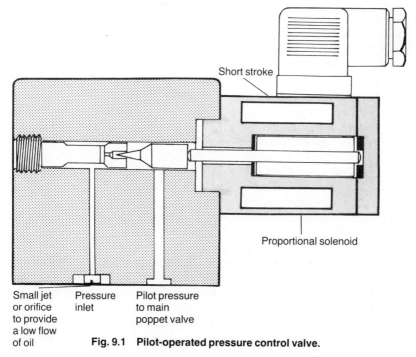

Short stroke

Proportional solenoid

Small jet   Pressure   Pilot pressure
or orifice   inlet   to main
to provide   poppet valve
a low flow
of oil   **Fig. 9.1   Pilot-operated pressure control valve.**

control valve it is necessary to reduce the effective area of the poppet, so that the force due to the maximum pressure acting upon the area of the poppet seat is lower than the maximum solenoid force.

For example, say a pressure of 315 bar (4500 psi) acts on a poppet with a seating area of $0.018\,cm^2$ ($0.001\,in^2$), the force would be 55 N (12 lbf). This is less than, say, the 80 N (18 lbf) operating thrust from a proportional solenoid. The maximum flow in this case would be very small, approximately 2 l/min (0.5 imp. gal/min). This is unsuitable for normal pressure control applications.

However, this small flow when used as a pilot function is capable of controlling a larger valve with a typical flow rate of 450 l/min (100 imp gal/min) or even more.

9.2.2 Flow Control Valves

The pressure-compensated flow control valve was developed into an electrically modulated valve almost two decades ago. An electrical servo motor was attached by gears to the throttling device and gave an output that was proportional to input signal.

This valve was a progression from the manual, flow controlled valve. Gears are incorporated as the drive mechanism to reduce the torque to suit the motor.

Present proportional flow control valves use a solenoid to produce a rotary action. A high torque involute cam is fitted between the solenoid and the throttling device to produce a rotary action and subsequently a variable flow output.

Sliding spool valves are also used as flow control valves especially when the spool is manipulated by a solenoid. Most spool type proportional valves are in fact directional control valves which use only one solenoid.

9.2.3 Directional Control Valves

The directional control valve operated by proportional solenoids can give a variable flow in one of two directions. Directional valves which have spool movements of only 1 mm (0.04 in) will not allow a large flow of oil through them, but like pressure control valves, they can be used to pilot a larger size of valve.

Various manufacturers of proportional solenoids use the current load system similar to that used for servo valves (see chapter 10). The solenoids need to produce a large stable force from the current available.

In all cases feedback of current is included in the electronic control circuit to stabilize the output force of the solenoid. This is because the impedance in the solenoid coil is altered owing to heat. A typical current drawn for this type of solenoid is 700 mA, to give a maximum force output.

Usually the pilot valve is a pressure-reducing valve (Fig. 9.2). The pressure increases to one of the pilots of the main spool and offsets this against the spring, creating flow through the main spool valve. The accuracy, therefore, is dependent upon the uniformity of the spring compression.

As the stroke of the solenoid is short, there are added problems for manufacturers endeavoring to provide suitable valve spools. To maintain constant flow rates from this type of valve independent of pressure and temperature variations, a compensator spool must be incorporated into the valve construction.

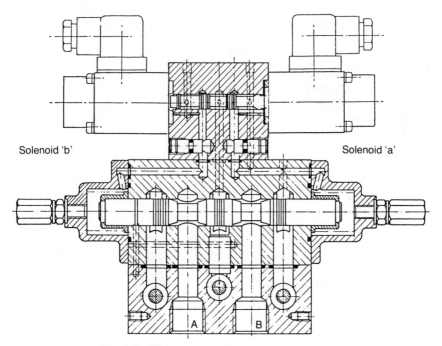

Solenoid 'b' Solenoid 'a'

Fig. 9.2 Pilot-operated directional control valve.

9.2.4 Dither

To overcome the sticking effect of the armature, a dither signal is loaded on to the current so that the solenoid moves at all times. The frequency is about 2–3 kHz. Figure 9.3 shows how this is done.

It is customary to provide an alternating signal to one of the amplifiers which control the solenoid. The desired value is given usually to the positive pin and the dither or alternating signal to the other pin of the amplifier. The output is now a modulating value with respect to the input signal. The effect is to oscillate the solenoid and, in turn, the valve spool. Sticking of the solenoid and the spool in the valve body is prevented (Fig. 9.4).

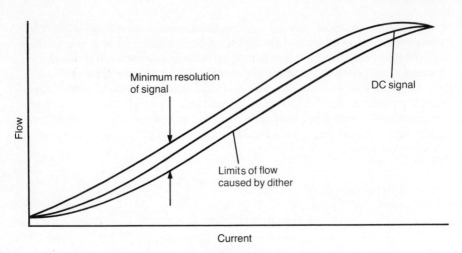

Fig. 9.3   Input current to output flow graph showing the effect of dither.

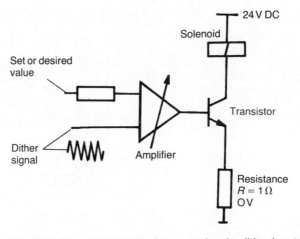

Fig. 9.4   Part of the control electronics incorporating the dither input signal.

9.3   LONG STROKE SOLENOID VALVES

A more advanced form of operation for proportional control valves is to use long stroke solenoids. To describe this type of solenoid control it is necessary to understand the mode of operation.

The voltage which is controlling the solenoid is produced by 'pulse width modulation', PWM (see appendix G). Early types of voltage controlled valves used a push/pull method of controlling the solenoid but this is no longer so. It is normal now for long stroke valves to have current feedback and this is used to generate the PWM voltage input.

The current drawn by long stroke solenoids can be as high has 3 A. However, this current is capable of producing a much greater force from the solenoid. The stroke of this type of solenoid is between 3.5 and 5 mm (0.12–0.2 in). It is possible to have valves operated by this type of solenoid capable of passing flow rates of 80 l/min (18 imp. gal/min).

9.3.1 Long Stroke or Voltage Controlled Valves

Long stroke solenoid valves work in a completely different way from current controlled valves in that they rely on an external sensor to produce their accuracy. This means that the accuracy of the valve does not rely on the linearity and characteristics of a solenoid. The spool movement acts against a spring and therefore the force of the solenoid is in proportion to the distance moved.

The external position sensor produces a repeatability of flow from the valve which is better than 1%.

Currently, the manufacturers who use voltage controlled valves use a linear variable differential transformer (LVDT) as the position sensor.

9.3.2 Position-sensing of the Spool

The LVDT is a simple transformer in which the primary windings are supplied with an alternating current of approximately 10 kHz. As the core is moved by the solenoid, it affects the induced voltage produced in the secondary windings. A decoder and supply circuit are needed to produce a varying DC voltage output, which can be compared with the set-point value.

Since the long stroke solenoid (Fig. 9.5) was developed it has had a great impact in hydraulic applications. Its features are that it:

1. is very accurate;
2. is voltage controlled;
3. has low hysteresis of signal;
4. has external feedback;
5. is direct acting.

9.4 VALVE SPOOL DESIGN

The valve spool used for proportional valves is altered in order to achieve a linear flow charactersitic. This is usually done with notches on the spool lands (Fig. 9.6). These are produced by either milling a semicircular notch in to the spool lands or coining a shaped notch with a press.

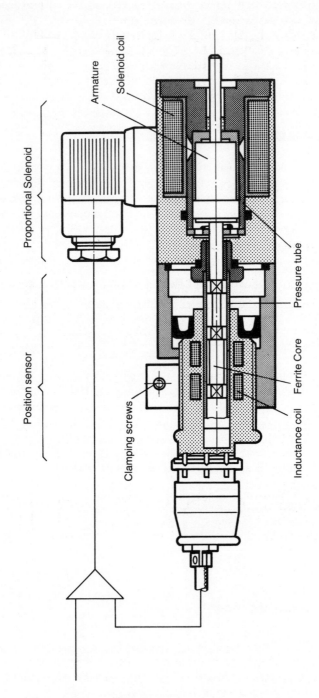

Fig. 9.5   Long stroke proportional directional valve.

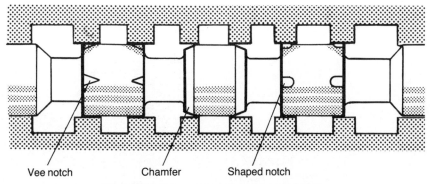

Vee notch   Chamfer   Shaped notch

Fig. 9.6   Various methods of spool notching.

9.5   PRESSURE COMPENSATION

To provide stability of flow from a flow control or directional control valve, a temperature and load compensator must be fitted to the input of the proportional valve. This compensator is a pressure drop device which holds the differential between the input and output pressure constant (Fig. 9.7).

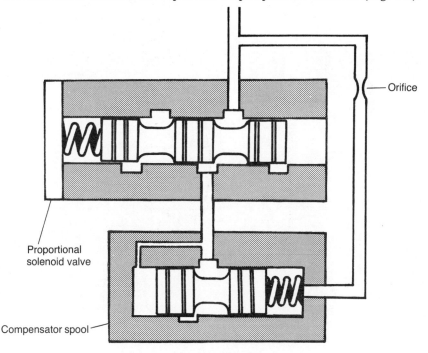

Fig. 9.7   Pressure compensation.

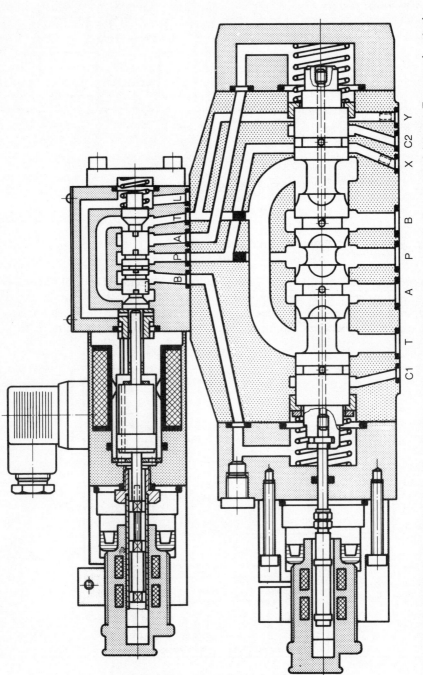

C1   T   A   P   B   X   C2   Y

Fig. 9.8 Pilot-operated spools showing the load compensator holes. A and B, actuator ports; P, pressure inlet port; T, reservoir or tank port; X, external pilot pressure port; Y, external pilot drain or tank port; C1 and C2 extra ports to accommodate pressure and temperature compensation.

When the temperature of the oil rises, the viscosity drops. The pressure at the inlet becomes lower so the compensator closes to the new setting, thus maintaining a constant flow.

A similar action occurs when the load pressure varies. The load pressure on the outlet port of the proportional valve increases and is sensed on the spring side of the compensator. The inlet pressure is then equal to the sum of the load pressure and the spring force. Directional valves must provide feedback pressure from each of the load or cylinder ports to control the compensator.

Pilot-operated spool valves, of port diameter between 16 and 25 mm, are designed with extra ports to accommodate a single pressure and temperature compensator valve (Fig. 9.8).

9.6 FREQUENCY RESPONSE

The frequency response of proportional control valves used to be very low, approximately 3–5 Hz. However, valves are now available which will respond to input signals of 70–80 Hz.

The consistent way of comparing any servo system for response is to use a bode diagram (Fig. 9.9). This diagram is constructed from two graphs imposed upon each other. The graphs have three abscissa, the base given as a logarithmic scale, the vertical axis in degrees and decibels. One graph shows the angle of phase shift or delay, the other shows the amplitude ratio.

The graphs plotted on the bode diagram (Fig. 9.9) shows the frequency against (a) the ratio between input and output values, when compared on a

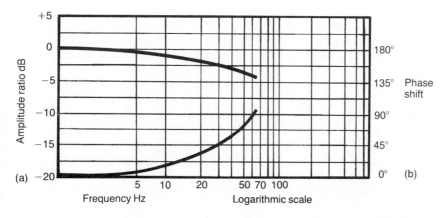

Fig. 9.9   Bode diagram showing frequency response and phase shift with amplitude ratio.

logarithmic scale, and (b) the delay that the output signal produces when compared to the input or set-point value.

The amplitude ratio on a logarithmic scale is given by:

$$dB = 20 \times \log_{10} \frac{\text{output voltage}}{\text{input voltage}}.$$

The angle of phase lag is produced by expressing the time delay between the input and output signals, at any given frequency, as the number of degrees the input signal will move through one cycle.

EXAMPLE: Phase shift: a delay of 2 ms in a 50 Hz frequency is 3·6°.

$$\frac{\text{frequency } 50 \text{ Hz} \times \text{delay } 2 \text{ ms} \times 360°}{\text{time } 1000 \text{ ms}} = 3·6°$$

Amplitude ratio: input signal of 10 V DC, output voltage 8 V DC

$$\frac{8}{10} = 0.8, \log 0.8 \times 20 = -0.0969 \times 20$$
$$= -1.938 \, dB$$

It is generally accepted that a servo system will give a delay of 45° at the maximum frequency of oscillation and the amplitude ratio would be better than −3 dB. This means that the output of a valve would be 70% of the input value and with a time delay of 1.56 ms at a frequency of 80 Hz.

9.7 AMPLIFIERS FOR PROPORTIONAL SOLENOIDS

Proportional valve amplifiers are designed to match the features of the valve with which they are used. Therefore they cannot be interchanged with any other valve. The short stroke valve takes constant current to the solenoid. The amplifier card has provision for an internal feedback signal from the coil to maintain the current output, regardless of any variation of the impedance of the solenoid coil.

The amplifier for the long stroke solenoid always compares the set-point or input voltage with the actual value of the external feedback transducer. On amplifiers for long and short stroke solenoids there are usually trim potentiometers which alter the gain to the valve and allow the valve to be set up correctly.

Output transistors are used on the amplifier card to switch the power needed to operate the solenoid. The input signal can be as low as 2.5 mA and solenoid currents up to 2.5 A. This will give a power gain on the amplifier card of approximately 1000:1. Any fluctuation in the input signal due to electrical noise (see chapter 7) will be amplified by the value of this power gain.

9.7.1 Ramp Signals

To increase the output of a valve slowly from the minimum setting to the desired setting requires a ramp signal. Delay times of 0.1–5 seconds are usually provided for proportional valves. The delay to desired value ratio is normally uniform in nature.

This ramp function is used to provide acceleration and deceleration rates for the load movement.

QUESTIONS

1. Name the various types of proportional valve.

2. Give the difference between long and short stroke solenoids.

3. What is dither and why it is used?

4. What is PWM and what is it used for?

5. Explain why pressure compensation is needed with proportional valves.

6. What are LVDTs used for with proportional valves?

10

Hydraulic Servo Valves

Servo valves were first introduced to control hydraulic systems nearly 50 years ago. The superiority of their design has withstood the time gap since their inception in that very few changes have been made. This chapter gives an overview of servo valves. They are noted for their high degree of controllability and fast response; they are not used in pneumatic systems.

10.1 INTRODUCTION

The most common type of servo valve that is manufactured is the spool type. This valve operates by means of a spool sliding in a body under the influence of hydraulic pressure. A torque motor is used to create this pilot pressure. The principle of the servo valve is to amplify a small electrical signal and convert it into hydraulic pressure or a flow of oil.

Spool type servo valves are not the only type available. They are also constructed in the form of plate or rotary valves. This chapter discusses the spool type valve because it is the one with which the engineer will most frequently come into contact.

10.1.1 Description of the Torque Motor

Servo valves are current-controlled in a similar way to short stroke proportional valves. The electrical signal operates a torque motor which has a flapper attached to it (Fig. 10.1). The torque motor is made up from two electromagnetic coils which are placed in a permanent magnetic field. The armature is supported by a flexible tube. This tube separates the oil from the electrical components. Attached to the armature is a flapper which terminates in a slender feedback wire. This wire is also attached to the spool.

10.1.2 Producing the Spool Movement

The servo valve operates on the pressure drop principle. Oil under pressure flows through two jet pipes or nozzles. The flapper is centralized between

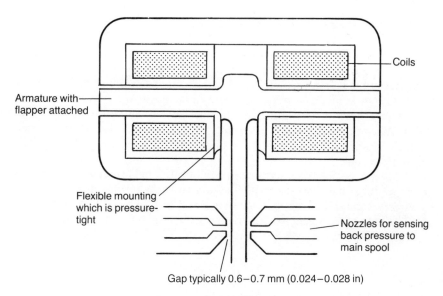

Gap typically 0.6–0.7 mm (0.024–0.028 in)

Fig. 10.1   The torque motor.

these nozzles and the flowing oil returns to the reservoir (Fig. 10.2). As the coil moves, under the influence of the current load, the flapper moves closer to one of the nozzles, causing a pressure build-up on that side.

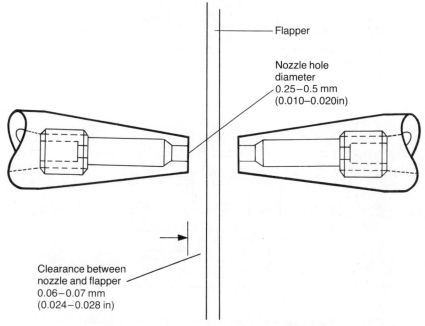

Fig. 10.2   Cross section of the pressure drop device.

10.1.3 Feedback of the Spool Movement

Figure 10.3 shows the servo valve at rest. As the torque motor is energized the pressure in nozzle 'a' increases (Fig. 10.4); the spool moves under the influence of the rise in pressure which moves the end of the slender feedback wire. The movement is in the opposite direction to the movement of the flapper and therefore a bending action takes place in the feedback wire (Fig. 10.5). When the tension in the feedback wire reacts with the force produced by the armature, the flapper is centralized and the pressure build-up it has created is reduced.

The valve spool now remains in equilibrium, held there by the tension in the feedback wire and the magnetic force from the torque motor. Oil then flows across the spool from ports P to A and B to T. The oil flow is in direct relationship to the current load on the torque motor.

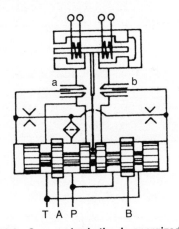

Fig. 10.3   Servo valve in the de-energized state.

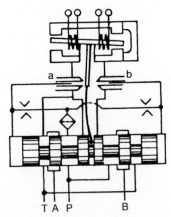

Fig. 10.4   Energization of the torque motor which increases the pressure to the 'a' pilot. The feedback wire is bending under load.

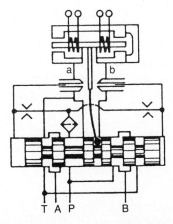

Fig. 10.5 The spool has now moved, centralizing the flapper. The tension in the feedback wire is equal to the energization force of the torque motor.

10.2 THROTTLING ACTION

As was stated in chapter 2, oil flowing through a throttling device creates a pressure drop. With servo valves this pressure drop is limited, by calculation during the system design, to 70 bar (1000 psi). It is necessary, therefore, to add at least 50% extra to the pressure required to move the load (called load pressure) to compensate for the high pressure drops encountered with servo valves. This excess pressure is lost in heat.

Because the spool movement is very small, high oil flow velocities occur across the spool. This can cause an erosion of the valve and so shorten its life.

Servo valves are extremely accurate, and should be reserved for systems where accuracy is of primary importance.

10.3 NEED FOR THE WHOLE SYSTEM TO HAVE STIFFNESS

Oil compresses very slighly when under pressure (see chapter 1). However, the effect that compressibility has on a servo system is important for the positional accuracy of the actuator.

The whole hydraulic system can be considered as a large spring. The total compression can be calculated from the total volume and the total loads acting on the actuator. This then allows the natural frequency of the hydraulic system to be calculated.

To prevent oscillation of the load and instability of the whole system, the maximum operating frequency is limited to one-third that of the natural frequency. The frequency response of the servo valve must be better than the maximum operating frequency.

It is important to have a very accurate feedback sensor that continously monitors the actual position of the load. Stiffness in the transfer arms is needed because of the high inertia forces experienced with the movement of the system. This feature is not a fault and assists in maintaining accuracy of position at all times.

10.4 HIGH FREQUENCY RESPONSE

As the frequency of the input signal increases, the dynamic response of the valve tends to lag behind the input signal and produces a delay or phase shift. As the switching frequency increases the oil has less time to build up pressure against the load to its maximum value and so the amplitude of the pressure response decreases (Fig. 10.6).

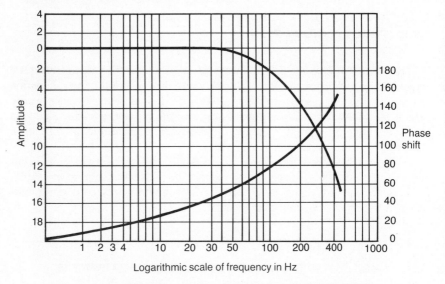

Logarithmic scale of frequency in Hz

Fig. 10.6 Bode diagram showing phase shift and amplitude ratio.

It was seen in chapter 9, when considering frequency response, that a servo valve must maintain a very close relationship between the input values and the output values. This is achieved by careful selection of the components which form the servo loop.

Obviously the more feedback loops placed around the various components, the higher degree of resolution there is. Servo valves should be reserved for systems which need this high degree of signal resolution, such as sophisticated robotics, aerospace, armaments, and some machine tools.

A special test device (Fig. 10.7) is used to check the response of the servo valve. The valve is mounted directly to the actuator so that the expansion of the piping is kept to a minimum.

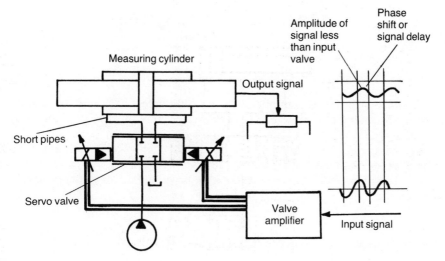

Fig. 10.7   Circuit showing the principle of the testing device.

10.5   CLEANLINESS OF THE OIL

A disadvantage of servo valves is that dirt causes severe damage to the valve during operation. A high degree of filtration is needed to ensure trouble-free operation.

The hole diameter of the nozzles can be as small as 0.25 mm (0.01 in). They can block with particles of dirt or silt up, unless great care is taken initially.

QUESTIONS

1.   What is a servo valve?

2.   Compare the characteristics of a servo valve with a proportional valve as discussed in chapter 9.

11

Design Examples

This chapter provides a number of worked examples. Each problem can be solved in a number of ways and each answer given is only one such solution. Consideration of these examples will provide a good foundation for using microcomputers with fluid power. Further benefit would be obtained by altering the examples and reworking them.

It will be seen that the structure of programs follows a recognizable pattern.

11.1 EXAMPLE 1: LOADING A HYDRAULIC PUMP

A hydraulic pump has to be loaded under microcomputer control. Later the pump has to be unloaded.

11.1.1 Tropics for Discussion

1. (a) Explain what is meant by 'unloading' of a pump and why it is done.
    (b) Produce a diagram showing the hydraulic components necessary.
2. Give a diagram showing the interfacing components between a port of the microcomputer and the hydraulic elements. Suggest typical voltage ratings.
3. Stating any assumptions necessary as to choice of microcomputer and I/O arrangements, give typical programming instructions both in machine code and assembly language and also in BASIC for:
    (a) loading the pump;
    (b) unloading the pump.

11.1.2 Possible Solution

1. (a) Unloading means taking the pressure away from the pump output. It is done to conserve energy and prevent overheating of the oil, when not doing any work.
    (b) The components diagram is shown in Fig. 11.1.

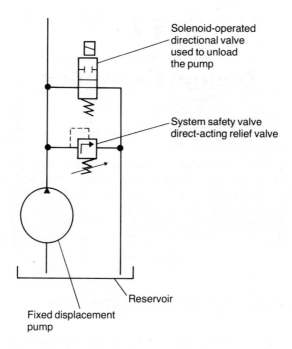

Solenoid-operated
directional valve
used to unload
the pump

System safety valve
direct-acting relief valve

Reservoir

Fixed displacement
pump

Fig. 11.1   Circuit for loading a pump.

2.   The valve solenoid is connected to a power supply (Fig. 11.2) through a solid state switching interface (see section 6.2.6).
Typical solenoid voltages: 220 V 50 Hz
                                                      110 V 50 Hz
                                                      24 V DC
A solid state relay could be used in each case.

3.   Assumptions: Z80 microprocessor used with PIO configured as in appendix D.

   (a)   Loading the pump.

        Machine code: Assembly language

        3EOF   LDA, OF   ;Set   CR A as output.
        D381   OUT(81),A

        3E04   LDA,04   ;Set bit 2 by writing to the whole port.
                                OUT(80),A

        BASIC
        10   OUT(81H), OFH   ;H indicates hexadecimal.
        20   OUT(80H), 04H

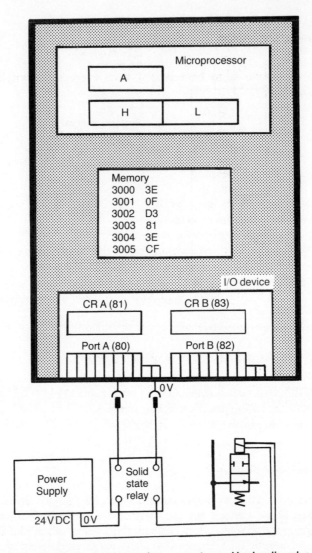

Fig. 11.2   Interfacing between microcomputer and hydraulic valve.

(b)   Unloading the pump. Assuming that the PIO needs only to be set-up once:

3E00	LDA,00	;Reset bit 2 by clearing whole port.
D380	OUT(80),A	

BASIC
70   OUT(80H),00H

11.2 EXAMPLE 2: AUTOMATIC UNLOADING OF A HYDRAULIC PUMP

The previous example is to be repeated with a pressure switch in the hydraulic circuit.

11.2.1 Topics for Discussion

1. Give reasons for the inclusion of the pressure switch.
2. Give a possible diagram showing:
    (a) the hydraulic components necessary;
    (b) the ports of a microcomputer and the interfacing components between the microcomputer and the hydraulic system.
3. Develop a programming flow chart for the following operation. Load the pump, wait until the pressure switch is made and then unload the pump. Ignore switch bounce.
4. Stating any assumptions made, convert the flow chart into a program in:
    (a) assembly language;
    (b) BASIC.
5. Explain what switch bounce is and how the effects can be eliminated by programming.

11.2.2 Possible Solution

1. A pressure switch is incorporated into a hydraulic circuit to indicate the reaching of the desired pressure within the system. It initiates the switching of the solenoid to unload the system when the required pressure is reached. To ensure that the pressure in the system does not decay when the unloading solenoid is energized, a check valve must be fitted before the pressure switch and after the unloading valve and relief valve.
2. A suggested layout of the hydraulic system and interfacing components is given, showing:
    (a) hydraulic circuit with interfacing to ports of the microcomputer (Fig. 11.3);
    (b) bit 0 of port A switches the solenoid on through a solid state switching device. Bit 0 of port B reads the pressure switch (Fig. 11.4).
3. The programming structure is shown in Fig. 11.5.

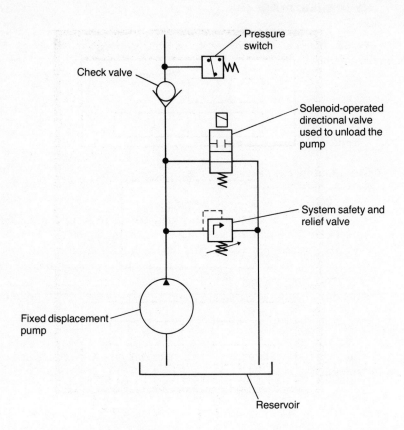

Fig. 11.3   Hydraulic circuit with pressure switch.

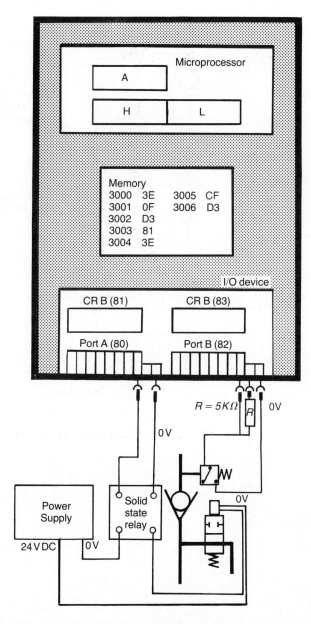

Fig. 11.4   Interface between microcomputer and control elements.

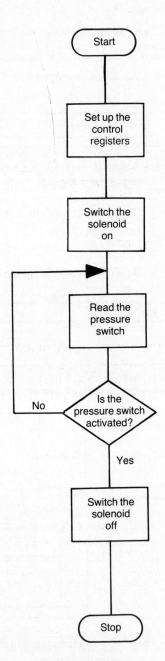

Fig. 11.5   Flow chart.

4. Assumptions made; the microprocessor used in the Z80 with PIO configured as in appendix D.

   (a) Loading the pump.

   Assembly language

   ```
   LDA,OF        ;Set CR A as output.
   OUT (81),A

   LDA,4F        ;Set CR B as input.
   OUT (83),A

   LDA,01        ;Set bit 0, port A using whole port.
   OUT (80),A    Switch on solenoid.

   IN A,(82)     ;Test pressure switch. See apppendix E.
   BIT 0,A
   JP NZ,DIS

   LDA,00        ;Reset bit 0, port A, using whole port.

   OUT (80),A
   ```

   ```
   BASIC
   10  OUT (81H), OFH   ; Port A output
   20  OUT (83H), 4FH   ; Port B input
   30  OUT (80H), 01H   ; Solenoid on
   40  IF INP (81H) AND 1 = 0 THEN GOTO 40 ; Wait for pressure
                                             switch to be
                                             activated (see
                                             appendix F).
   50  OUT (80H), 00H   ; Solenoid off
   ```

5. Controlling switch bounce. Switch bounce causes repeated changes of state in the signal when a switch is activated which may last for several milliseconds. Because the microcomputer can read the closing action in microseconds, it can read all these bounces giving the impression that the switch has been operated several times. To give positive control it must be certain that the switch has been operated, before processing the signal.

   One method is to read the switch constantly until the bounce has finished, Fig. 11.6. An alternative is to use a delay subroutine. Such a subroutine is given as part of the solution to example 6, Fig. 11.7.

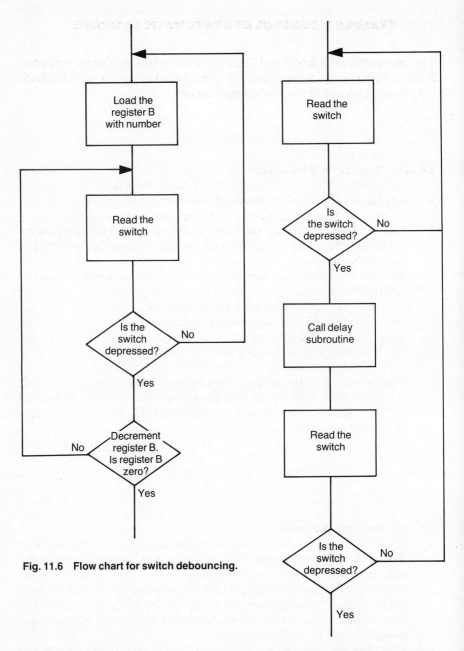

Fig. 11.6   Flow chart for switch debouncing.

Fig. 11.7   Alternative flow chart.

11.3 EXAMPLE 3: CONTROL OF A PNEUMATIC CYLINDER

Two pneumatic cylinders A and B are to be controlled by a microcomputer. Each cylinder is to have a pair of limit switches to provide feedback concerning the limits of travel of each piston rod.

11.3.1. Topics for Discussion

1. Explain what is meant by 'feedback' and why it is necessary.
2. Draw a possible pneumatic circuit showing the necessary valves.
3. Give a diagram showing the likely electrical interfacing components connecting the pneumatic circuit and the limit switches to the ports of a microcomputer.
4. Develop a programming flow chart given that the sequence of events is to be:
    (a) 'A' out until limit switch LS1 is contacted;
    (b) 'B' out until limit switch LS2 is contacted;
    (c) 'B' in until limit switch LS3 is contacted;
    (d) 'A' in until limit switch LS4 is contacted;
    (e) repeat the sequence.
5. Stating any assumptions made, develop and write a program covering the sequence 4(a) above. Include the setting up of any I/O device and allow for switch bounce. Give the program in
    (a) assembly language;
    (b) BASIC.

11.3.2 Possible Solution

1. Feedback provides information regarding the behavior of a system for the control unit to act upon.
2. The pneumatic circuit (Fig. 11.8) has two solenoid-operated directional control valves interconnected to the cylinders. With this system it is only necessary to switch on the appropriate solenoid to make the actuator move in the desired direction.
3. The interface components are solid state relays interfaced with a power supply to the solenoids. Port A is configured as input and Port B as output to the switches (Fig. 11.9).
4. The flow chart giving the sequence of operation is shown in Fig. 11.10.

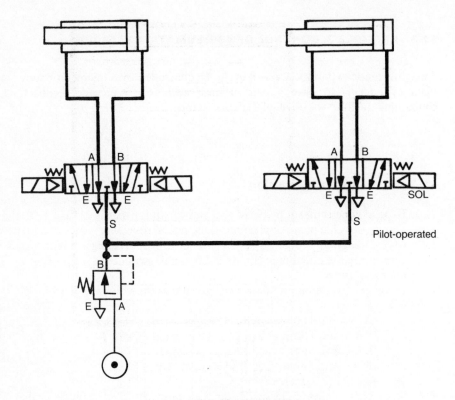

Fig. 11.8   Pneumatic circuit.

5.   Assumptions: the microprocessor used is the Z80 with PIO configured
     as in appendix D.
     (a)   Write the program in assembly language:

          LDA,OF   ;Set CR B as output.
          OUT (83),A

          LDA,4F   ;Set CR A as input.
          OUT (81),A

          IN A, (82)   ;Set bit 3, see appendix E.
          SET 3,A   Switch on solenoid S10.
          OUT(82),A

          LDB,FF   ;Load register B with number of
                          times the switch is to be checked.

          IN A, (80)   ;Read the switch port A bit 7.
          BIT 7,A

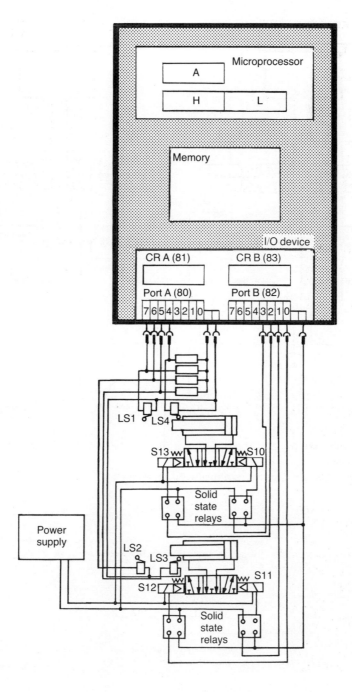

Fig. 11.9 Interfacing elements.

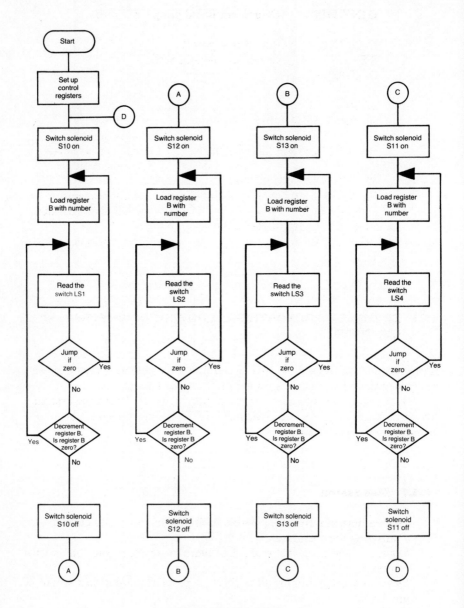

Fig. 11.10   Flow chart.

```
        JRZ,DIS

        DJNZ,DIS      ;Decrement B and jump if not zero.

        IN A(82)      ;Reset bit 3, port A,
        RES 3,A        switch off solenoid S10
        OUT(82),A
```

(b) BASIC

10	OUT(83H), OFH	;Port B ouput
20	OUT(81H), 4FH	;Port A input
30	LET A=INP(82H) OR 08H	;Switch on solenoid S10, port B bit 3
40	OUT(82H),A	
50	IF INP(80H) AND 80H=0 THEN GOTO 50	;Wait for pressure switch to be activated
60	LET A=INP(82H) and F7H	
70	OUT(82H),A	;Switch off solenoid S10
80	END	

11.4 EXAMPLE 4: PROPORTIONAL CONTROL OF A HYDRAULIC CYLINDER

A proportional valve is to be used with a hydraulic cylinder. It is to move a heavy load smoothly, in both forward and backward directions.

The piston rod is to move between limit switches. When the first limit switch is contacted the speed is reduced by 75%. On contacting the second limit switch the load is halted.

11.4.1 Discussion

1. Draw a diagram showing possible hydraulic circuitry and elements for controlling the operations.
2. Suggest interfacing between the hydraulic system and the micro-computer.
3. Develop a programming flow chart to provide the outward stroke of movement.
4. Assuming the acceleration and deceleration rates are set by the proportional valve control card, write a program for extending the cylinder rod, in:
    (a)   machine code;
    (b)   BASIC.

11.4.2   Possible Solution

1.   This application will use a proportional flow control valve with
     temperature and load compensation (Fig. 11.11).

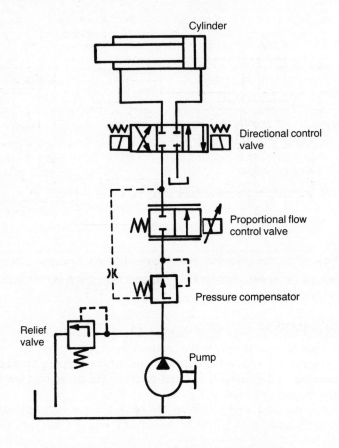

Fig. 11.11   Hydraulic circuit with proportional valve.

2.   A DC power supply produces 24 V DC for the proportional amplifier
     and solid state switches. A D/A converter is interfaced to port B. Port A
     has mixed inputs and outputs. Bits 7–4 read the limit switches, and bits
     3 and 2 the outputs to the solid state relays for the solenoids (Fig.
     11.12).
3.   The programming flow chart is shown in Fig. 11.13.

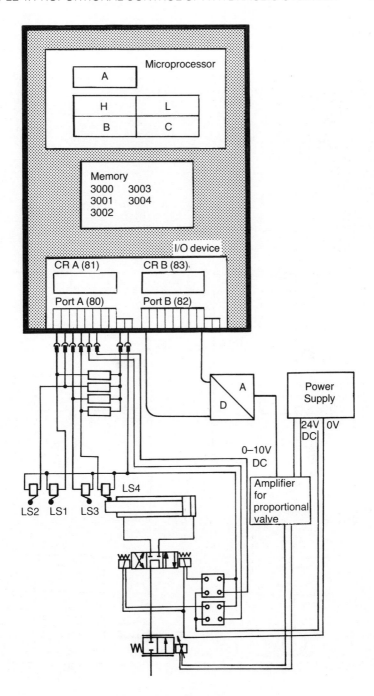

Fig. 11.12   Interfacing elements.

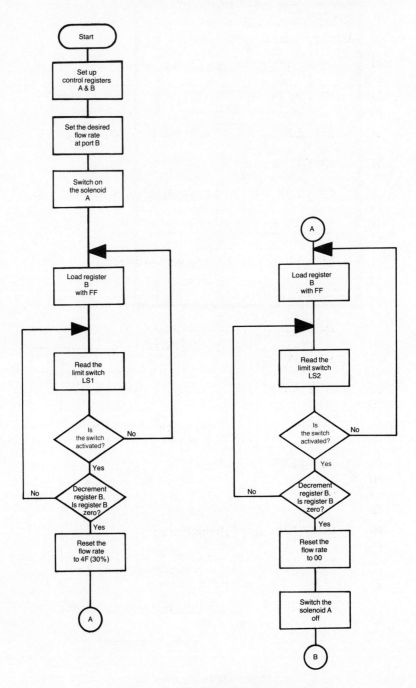

Fig. 11.13   Sequence of operation.

4.   Assumptions: the microprocessor used is the Z80 with PIO configured as in appendix D.

(a)   Machine code.

3E OF	LDA,OF	;Set CR B as output
D3 83	OUT(83),A	
3E CF	LDA,CF	;Set CR A as mixed input/output.
D3 81	OUT(81),A	
3E F0	LDA,F0	;Bits7–4 all inputs, 3–0 all outputs.
D3 81	OUT(81),A	
3E FF	LDA,FF	;Set proportional valve.
D3 82	OUT(82),A	
DB 80	IN A,(80)	;Switch Solenoid A on see appendix E.
CB DF	SET 3,A	
D3 80	OUT(80),A	
06 FF	LDB,FF	;Load register B with FF.
DB 80	IN A,(80)	;Read LS1.
CB 7F	BIT 7,A	
20 F8	JRNZ,DIS	;Jump if not zero.
10 F8	DJNZ,DIS	;Decrement B and jump if not zero.
3E 4F	LDA,4F	;Reduce value of proportional valve.
DB 82	OUT(82),A	

06	LDB,FF	;Load register D with FF.
FF		
DB	IN A,(80)	;Read LS2.
80		
CB	BIT 6,A	
77		
20	JRNZ,DIS	
F8		
10	DJNZ,DIS	;Decrement B and jump if not zero.
F8		
3E	LDA,00	
00		
DB	OUT(82),A	
82		
DB	IN A,(80)	;Switch solenoid A off.
80		
CB	RES 3,A	
9F		
D3	OUT (80),A	
80		

(b)   BASIC

```
 10  OUT(83H),OFH         ;Set up I/O
                          H indicates hexadecimal.
 20  OUT(81H),CFH
 30  OUT(81H),FOH
 40  OUT(82H),FFH         ;Set value to valve.
 50  LET A=INP(80) OR 08H ;Switch on solenoid A.
 60  OUT(82H),A
 70  IF INP(80) AND 80H=0 THEN GOTO 70 ;Wait for LS1
                                        to make
                                        contact.
 80  OUT(82H),40H         ;Reduce value of proportional
                           valve to 25%.
 90  IF INP(80H) AND 40H=0 THEN GOTO 90 ;Wait for
                                         LS2 to
                                         make
                                         contact.
100  OUT(82H),00H         ;Reduce value of proportional
                           valve to zero.
110  LET A=INP(80H) AND F7H ;Switch solenoid A off.
120  OUT(80H)A
130  END
```

11.5   EXAMPLE 5: CONTROL OF A PNEUMATIC CYLINDER WITH DIGITAL FEEDBACK

The position of the piston rod of a pneumatic cylinder is to be monitored by a microcomputer and a two-channel shaft encoder.

11.5.1   Topics for Discussion

1.   Stating any assumptions made, develop a program in machine code which increments a register for each pulse of the shaft encoder in one direction and decrements the register for the other direction. It is not necessary to show the interfacing of the pneumatic valve to the microcomputer as this example is just a small part of the complete system.
2.   Explain why machine code would be used with a shaft encoder in preference to a higher level language.

11.5.2   Possible Solution

1.   Assumptions made: the microprocessor used is the Z80 with PIO configured as in appendix D. The shaft encoder is connected to bits 0 and 1 of port A (Fig. 11.14). The flow chart (Fig. 11.15) shows how the reversing of direction of the drive shaft of the encoder is accommodated.
2.   High speed is necessary to ensure that all the pulses are monitored. BASIC would be too slow.

11.6   EXAMPLE 6: MICROCOMPUTER CONTROL OF A PROPORTIONAL VALVE

A microcomputer controls a proportional valve. The valve is to be controlled by a ramp input of voltage and then held steady. The ideal would be as shown in Fig. 11.16.

11.6.1   Topics for Discussion

1.   Stating any assumptions made, show how the ramp input might be achieved. Develop a part of a program to do it.
2.   Suggest a way of producing a voltage to match any shape of graph.

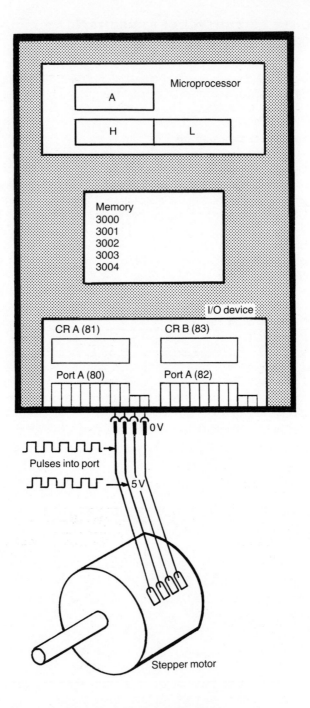

Fig. 11.14   Shaft encoder inputs.

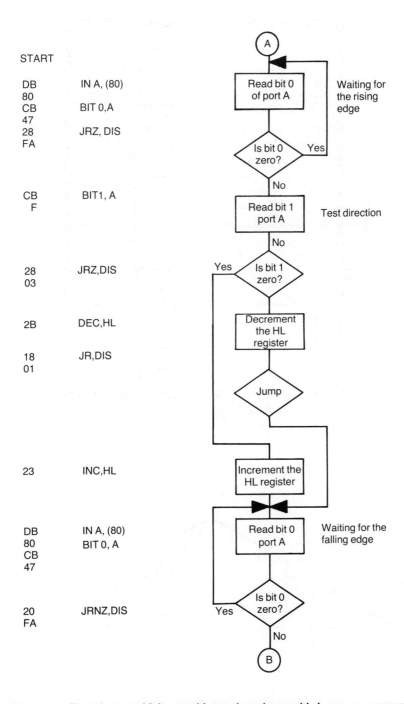

START

DB	IN A, (80)
80	
CB	BIT 0,A
47	
28	JRZ, DIS
FA	

| CB | BIT1, A |
| F | |

| 28 | JRZ,DIS |
| 03 | |

| 2B | DEC,HL |

| 18 | JR,DIS |
| 01 | |

| 23 | INC,HL |

DB	IN A, (80)
80	BIT 0, A
CB	
47	

| 20 | JRNZ,DIS |
| FA | |

Fig. 11.15   Flow chart combining machine code and assembly language program.

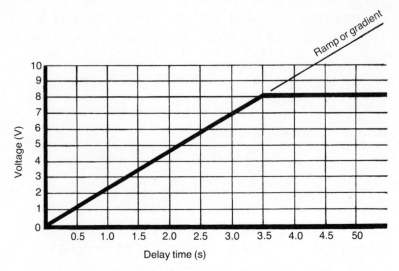

Fig. 11.16   Graph showing gradient of desired ramp signal.

11.6.2   Possible Solution

1.   Assumptions: a Z80 microprocessor is used with PIO as in appendix D. The ramp formed can only be produced by varying the time base to the increment (Fig. 11.17). The program shown in Fig. 11.18 is written in two separate parts. Firstly, the program is written to generate the increments or steps, then the program for the time delay is given.

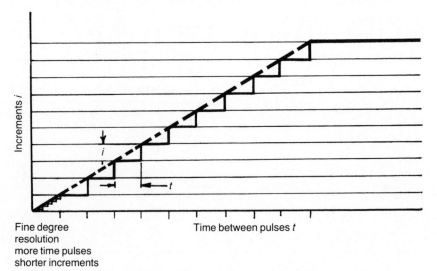

Fig. 11.17   Graph showing the time to increment ratio.

Program: increment steps

Machine code	Assembly language
OE 00	LDC,0
06 NN	LDB,NN
CD YY XX	CALL XXYY SUBROUTINE
OC	INC C
79	LDA,C
D3 80	OUT (80),A
10 F7	DJNZ,DIS

Program: delay subroutine T

XXYY	21 MM NN	LDHL,NNMM
	2B	DEC HL
	3E 00	LDA,00
	BC	CP H
	20 FA C9	JRNZ,DIS

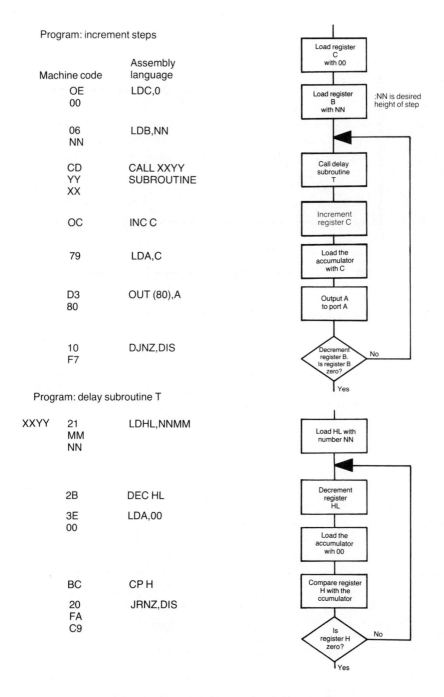

Fig. 11.18   **Programs giving increment and time delay.**

2.   Any voltage profile could be produced by prestoring values of i_0, i_1, i_2, etc. in memory and adding or subtracting as appropriate to produce the steps (Fig. 11.19).

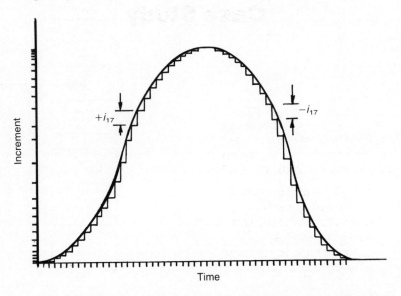

Fig. 11.19   Graph showing increment and time delay to generate a typical signal.

12

Case Study

This study concerns the design of a machine to perform hydromechanical deep drawing. It may be regarded as a typical example of fluid power with microprocessor control. The study needs to include the following:

1. description of the process;
2. mechanical requirements;
3. hydraulic control circuits;
4. feedback elements;
5. interfacing and input/output to microcomputer;
6. software requirements.

12.1 DESCRIPTION OF THE PROCESS

The machine is to be designed to perform hydromechanical deep drawing. A description of this process and how it needs to be carried out follows.

12.1.1 Hydromechanical Deep Drawing

Deep drawing is one of the most important processes in the production of sheet metal components, ranging from items for the motor industry to cooking utensils. Factors which limit the speed of the operation of the process and the geometrical variety of products are:

1. the quality and properties of materials;
2. the tools and their geometry;
3. lubrication between the touching surfaces of punch, workpiece and the die.

To eliminate these shortcomings, hydromechanical deep drawing was introduced.

157

12.1.2 Mode of Operation

The principle of the operation is similar to that of conventional deep drawing except that the cushion is replaced by an oil-filled pressure chamber.

The principle of operation is described below. A component blank is inserted into the press and the pressure plate is lowered. Limitation of workpiece size is necessary because of the very high forces created by the intensifier circuit.

The press cylinder is lowered until the former comes into contact with the workpiece (Fig. 12.1).

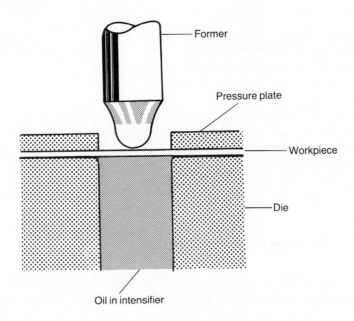

Fig. 12.1 Component workpiece ready for deep drawing.

The pressure in the intensifier is then raised to produce a reverse preforming action. This starts the forming of the workpiece around the minor diameters of the forming tool (Fig. 12.2). The press cylinder and the intensifier cylinder start the drawing operation. Variations are made to the pressure plate load during the operation.

As the difference in diameters between the die and the former reduces to a point where the metal would become creased, the intensifier pressure is reduced. This continues to support the metal of the workpiece but not to provide a reverse forming action (Fig. 12.3)

When the former is fully lowered it may be necessary to increase the pressure in the intensifier to maximum to produce an ironing effect upon the workpiece and 'set' the form.

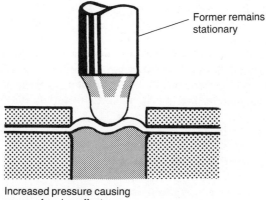

Former remains
stationary

Increased pressure causing
reverse forming effect

Fig. 12.2   Preforming action started.

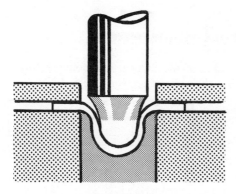

Fig. 12.3   Forming action.

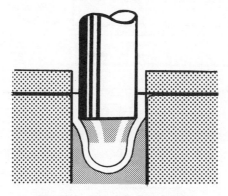

Fig. 12.4   Forming completed.

When the forming is completed (Fig. 12.4) the pressure in the intensifier is lowered to a safe level, the former is withdrawn and the component removed.

Reverse forming using a hollow former is also possible. Without hydromechanical forming a two-stage operation would be required. The complete shape can be formed in one operation, with the pressure of the oil in the pressure chamber forcing the material into the cavity in the core of the former.

12.1.3 Control Requirements

It is required to be able to deep draw specimens, remove and test them, and readjust the cycle until the desired component is achieved. Hence it is necessary to be able to store in memory the particular sequence of values in order to produce further components.

12.2 MECHANICAL REQUIREMENTS

A diagram of a layout of mechanical components and hydraulic cylinders necessary to perform the process is given in Fig. 12.5.

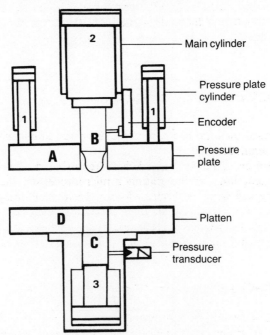

Fig. 12.5   Press structure.

12.2.1 Hydraulic Components

There are three distinct hydraulic components:

1. *Pressure plate cylinders* are two double-acting cylinders which operate the pressure plate (A) which is lowered to hold the workpiece. Pressure has to be varied as forming takes place. The force of the pressure plate is capable of withstanding the induced load created by the intensifier.
2. The *press cylinder* is a double-acting cylinder that drives the forming tool (B). The forming tool must be controlled for varying speed, position and thrust.
3. The *intensifier cylinder* intensifies the pressure in the pressure chamber (C). As the drawing takes place so the piston must move to maintain the correct pressure in the pressure chamber.

12.2.2 The Press Structure

A platten (D) is constructed into the framework of the press which provides a reactionary force for the load from all the cylinders. The pressure plate (A) is attached to the approach cylinders and passes around the press cylinder.

A holder (B) is attached to the press cylinder which holds the forming tool. To this holder is connected the linear encoder for position monitoring.

The intensifier (3) is fastened to the fixed platten (D) and creates a pressure of 700 bar in the cavity (C) on the underside of the die. A pressure transducer is fitted into the wall of the intensifier to monitor the hydraulic forming pressures.

12.3 HYDRAULIC CONTROL CIRCUITS

Hydromechanical forming needs variable pressures and speeds from all the three actions simultaneously.

This project could be regarded as requiring a slow-speed, servo-controlled system but, by integrating a microprocessor, the variable-gain control of a typical servo is dispensed with. Instead of using servo valves, with their inherent drawbacks of high cost and being easily damaged by impurities in the oil, proportional solenoid valves are used at half the cost and with less dependency on oil cleanliness.

12.3.1 Operating Procedure of the Hydraulic Circuit

With the pumps (6) and (7) running and the guard open, the operator fills the intensifier full of oil, loads a blank disc or workpiece and initiates the forming process by closing the guard (Fig. 12.6).

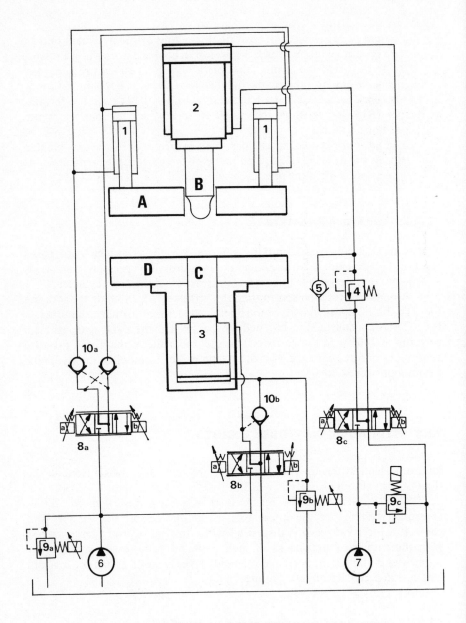

Fig. 12.6   A control circuit.

The first operation of the press is to lower the pressure plate onto the workpiece. To do this the load pressure is set on the proportional relief valve (9a). The pressure plate cylinders lower when a value is given to solenoid a, of the proportional directional control valve (8a).

The pressure relief valve (9c) is given a value and the press cylinder is lowered by a value being given to valve (8c). As the pressure increases in the annulus side of the cylinder owing to the force of the pressure on the full bore area, it is relieved by the relief valve (4). This relief valve acts as a counter-balance for the weight of parts of the press cylinder.

The position of the former is monitored by the linear encoder and the microcomputer. When the former reaches the workpiece, the pressure in the intensifier chamber is raised to the desired value.

The pressure in the intensifier is raised by giving a value of voltage to the relief valve (9b) and a value of voltage to the directional valve (8b). The control action of the microcomputer takes over the forming process by monitoring the pressure transducer and the linear encoder.

At each increment of the linear encoder all the values of the proportional valves are set with the pressure transducer constantly causing the adjustment of the intensifier cylinder speed.

When the pressure plate force reduces so that the pressure in the system to these cylinders is less than that required by the intensifier cylinder, the valve (8b) is de-energized. The load constant is maintained by the setting to the relief valve (9a).

When the forming action is completed, the pressure in the intensifier is reduced and the press cylinder is raised. This is accomplished by giving a value to solenoid b of valve (8c). Oil now flows through the check valve (4), to raise the press cylinder.

The pressure plate cylinders are raised by a value being given to solenoid b of the valve (8a); the intensifier ejects the formed workpiece by a value being given to the solenoid b of the valve (8c).

12.4 FEEDBACK ELEMENTS

In order that the microcomputer can monitor and record exactly what is happening, the following feedback elements are needed.

12.4.1 Linear or Shaft Encoder

The linear encoder is fastened to the platten (B) and the structure of the press cylinder to give the position of the former as the drawing takes place. It could alternatively be attached to the former (B) and the fixed platten (D) to give the exact measurement of the depth to which the component is drawn. Deformation in the structure of the press would then not be recorded.

12.4.2 Pressure Transducer

This is mounted into the high pressure side of the intensifier (3) or the pressure chamber (C).

The forming process demands a variable hydrostatic force (pressure) of up to 700 bar on the workpiece blank. The requirement is to vary this pressure as the forming progresses.

12.5 INTERFACING AND INPUT/OUTPUT TO MICROCOMPUTER

The following are required:

1. Data to six amplifier cards for the proportional valves. This can be done from one single port, using a second port to latch data to whichever proportional valve is required.
2. Data from the pressure transducer via an analog to digital converter.
3. Data from a linear encoder.

Figure 12.7 shows that at least three ports are required. The solution given is to use two PIOs.

The microprocessor used is the Z80 with two PIOs configured as in appendix D.

12.6 SOFTWARE REQUIREMENTS

During the cycle of operation there is need to set each of the proportional valves to a predetermined value. As the press cylinder moves, the predetermined hydrostatic pressure must be kept to within, say, 0.7% of the desired value.

It is therefore constantly necessary to monitor and adjust the intensifier cycle. The reaction time of the oil is approximately 7000 bar/s. The pressure must, therefore, be monitored every 200 μs.

The response of the valve to change is approximately 2% per millisecond. This arrangement raises the two issues which most frequently occur in control problems:

1. data acquisition, especially initially;
2. timing.

12.6.1 Data Acquisition

Once data have been acquired they can be stored in memory for use on other occasions. The main problem lies in acquiring the data initially. There are several ways of doing this.

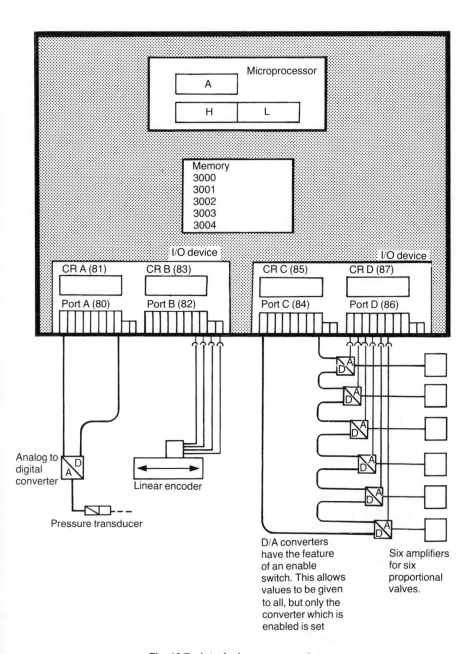

Fig. 12.7 Interfacing components.

1. Entering data manually; this may be done by filling in figures in a table by guesswork, perhaps based upon experience. As it would be tedious to type in some thousand figures, only a sample would be done. The microcomputer would complete the remainder by interpolation. The data can be amended subsequently to achieve optimum results by trial and error.

    The program for the manual entering of data can be done in BASIC.

2. Teach and learn; the well-known technique of steering a robot through its paces manually and then having the data stored in the memory for repeat performances is not applicable here, as it is difficult to control six proportional valves simultaneously by hand.

12.6.2 Timing

The question invariably arises as to whether the microcomputer can operate fast enough to do all that is required of it. Whilst drawing is being done the microprocessor must record every pulse from the linear encoder and at the same time provide new data to each proportional valve.

Consider the following parameters as typical: The forming tool moves at, say 20 mm/s, maximum speed; data are to be stored, say, at every 0.5 mm of movement, which is given by pulses from the linear encoder. The time between pulses is then 25 ms or 25 000 µs.

A typical average time for each machine code instruction is 2 µs. Patently there will be adequate time between recording pulses to service the valves, but only if machine code is used for this part of the program.

12.6.3 The Flow chart

The essential part of the program is the control of the press as it does the forming. This is a dynamic situation and must be done in machine code. A flow chart showing the logic for this part of the program only is given in Fig. 12.8. The following points should be noted.

1. The program has to be such that it constantly adjusts the pressure in the intensifier and tests whether the linear encoder has changed state, i.e. moved on a step.

2. The control data will be stored in memory in blocks of six pieces of information. At each position of the linear encoder, the settings to five of the proportional control valves are given. The intensifier pressure is constantly monitored by feedback from the pressure transducer and the remaining proportional valve is adjusted to maintain the desired pressure setting.

    The store of data is called a stack. A register in the micro-processor, designated the stack pointer, will hold the address of the part of the stack currently in use. The stack pointer will then be incremented in blocks of six.

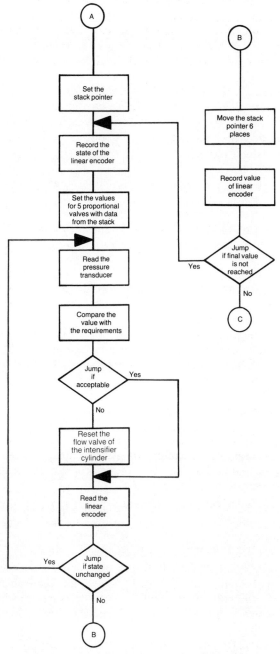

Fig. 12.8 Flow chart showing control of the press.

Appendix A

Hydraulic Symbols

Constant flow pump,
single outlet.

Constant flow pump,
double outlet.

Variable flow pump,
single outlet.

Variable flow pump,
dual direction of rotation.

Hydraulic motor,
single direction of rotation.

Hydraulic motor,
dual rotation.

Variable flow hydraulic motor,
single direction of rotation.

Hydraulic motor dual rotation,
variable flow.

Hydraulic pump or motor,
single flow.

Hydraulic pump or motor,
rotation one way only.

Dual rotation hydraulic
pump or motor.

Variable flow hydraulic
pump or motor.

Variable pump or motor, rotation one way only.

Variable pump or motor, dual rotation.

Check valve without spring.

Check valve with spring.

Pilot-operated check valve.

Pilot-operated check valve (with drain port).

Throttle check valve.

Shuttle valve.

Power source (air or oil).

Motor (denotes electric motor).

Motor (denotes i/c engine).

Drive shaft.

Coupling.

Power line.

Tank return line.

Pilot line.

Flexible hose.
Power line junction.

Crossing power lines (non-connecting).

Air bleed.

Test point.

Quick release coupling.

Rotating coupling (multi-line).

Reservoir (with baffle).

Accumulator.

Filter.

Cooler.

Heater.

Pressure gauge.

Flow meter.

Directional Valves

3 position valve symbol
(center box shows flow condition at rest).

Boxes show flow paths:
(a)   initiating action;
(b)   return action (O) spring center condition.

Complete valve symbols:

P, pump (pressure source),
T, tank (return line),
A, B, service ports,

L, leak oil port.

2/2 directional valve.

3/2 directional valve.

4/3 directional valve.

6/3 directional valve.

Operating Actuators

Hand lever.

Push button.

Foot pedal.

Cam follower.

Roller operator.

Spring return.

Spring centering.

Electrical solenoid (single-acting).

Electrical solenoid (double-acting).

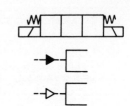

Hydraulic pilot.

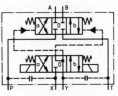

Pneumatic pilot.

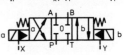

Pilot-operated valve
(diagram showing pilot lines and drain lines;
drawn usually like this in manufacturer's catalog).

Drawn symbol for pilot-operated valve
(double-acting).

Drawn symbol for single-acting valve.
(usual symbol shown on circuit diagrams).

Roller operated servo valve
(used as flow balancing valve).

Proportional valve.

Electro-hydraulic servo valve.

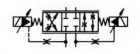

Pressure Valves

Poppet function (normally open).

Poppet function (normally closed).

Direct-acting pressure relief valve.
(fixed and variable adjustment).

Pilot-operated relief valve.

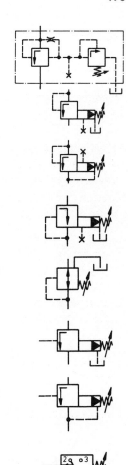

Pilot-operated relief valve
(external pilot, external drain).

Pilot-operated relief valve
(external pilot, internal drain).
Pressure reducing valve
(fixed and variable adjustment).

Pilot-operated pressure reducing valve.

Pressure reducing valve (three position).

Sequence valve
(external drain for constant load).

Sequence valve (external drain, load varies
with back-pressure in the tank line).

Pressure switch.

Throttle Valve

Throttle (fixed and variable adjustment).

2-way pressure compensated flow control.

3-way presure compensated flow control
(used as a priority flow controller, excess
flow returns to tank or secondary circuit).

Flow divider.

Hydrostatic transmission.

Compact symbol.

Rotary actuator.

Single-acting cylinder
(return action by load).

Single-acting cylinder
(spring return).

Double-acting cylinder.

Double rod cylinder.

Cylinder with cushions.

Cylinder with adjustable cushions.

Telescopic cylinder.

Intensifier cylinder.

Appendix B

Pneumatic Circuit Symbols

Double-acting cylinder with adjustable cushions.

Single-acting cylinder with spring return.

Double-acting cylinder.

Rotary actuator.

3/2 seated directional control valve.

3/2 directional control valve
(mechanical actuator and spring return).

3/2 directional control valve
(mechanical actuator with air pilot return).

3/2 directional control valve
(with mechanical and air pilot actuator).

3/2 roller and air pilot operated
directional control valve.

Pneumatic presetting timing valve.

5/2 directional control valve
(mechanical actuator with spring return).

3/2 directional control timer valve.

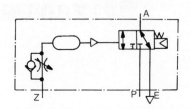

5/2 pneumatic pilot directional control
valve with detent.

5/3 pneumatic pilot directional control valve
with pneumatic actuator.

Center box showing flow paths: P to A and B.

Center box showing flow paths; A and B to exhaust.

3/2 solenoid pilot-operated directional control
valve with manual override and spring return.

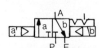

3/2 solenoid pilot-operated directional control
valve with detent.

Throttle valve.

Pressure regulator valve.

Air filter with water trap.

Oil lubricator.

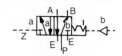

Appendix C

Z80 Instruction Set

Z80-CPU instructions sorted by Mnemonic

Obj code	Source statement	Obj code	Source statement	Obj code	Source Statement	Obj code	Source statement
BE	ADC A, (HL)	DD8605	ADD A, (IX + D)	CB59	BIT 3, C	EDA1	CPI
DD8E05	ADC A, (IX + d)	FD8605	ADD A, (IY + d)	CB5A	BIT 3, D	EDB1	CPIR
FD8E0F	ADC A, (IY + d)	87	ADD A, A	CB5B	BIT 3, E	2F	CPL
8F	ADC A, A	80	ADD A, B	CB5C	BIT 3, H	27	DAA
88	ADC A, B	81	ADD A, C	CB5D	BIT 3, L	35	DEC (HL)
89	ADC A, C	82	ADD A, D	CB66	BIT 4, (HL)	DD3505	DEC (IX+d)
8A	ADC A, D	83	ADD A, E	DDCB0566	BIT 4, (IX+d)	FD3505	DEC (IY+d)
8B	ADC A, E	84	ADD A, H	FDCB0566	BIT 4, (IY+d)	3D	DEC A
8C	ADC A, H	85	ADD A, L	CB67	BIT 4, A	05	DEC B
8D	ADC A, L	C620	ADD A, N	CB60	BIT 4, B	0B	DEC BC
CE20	ADC A, N	09	ADD HL, BC	CB61	BIT 4, C	0D	DEC C
ED4A	ADC HL, BC	19	ADD HL, DE	CB62	BIT 4, D	15	DEC D
ED5A	ADC HL, DE	29	ADD HL, HL	CB63	BIT 4, E	1B	DEC DE
ED6A	ADC HL, HL	39	ADD HL, SP	CB64	BIT 4, H	1D	DEC E
ED7A	ADC HL, SP	DD09	ADD IX, BC	CB65	BIT 4, L	25	DEC H
86	ADD A, (HL)	DD19	ADD IX, DE	CB6E	BIT 5, (HL)	28	DEC HL
		DD29	ADD IX, IX	DDC8056E	BIT 5, (IX+d)	DD28	DEC IX
		DD39	ADD IX, SP	FDCB056E	BIT 5, (IY+d)	FD28	DEC IY
		FD09	ADD IY, BC	CB6F	BIT 5, A	2D	DEC L
		FD19	ADD IY, DE	CB68	BIT 5, B	3B	DEC SP
		FD29	ADD IY, IY	CB69	BIT 5, C	F3	DI
		FD39	ADD IY, SP	CB6A	BIT 5, D	102E	DJNZ DIS
		A6	AND (HL)	CB6B	BIT 5, E	FB	EI
		DDA605	AND (IX+d)	CB6C	BIT 5, H	E3	EX (SP), HL
		FDA605	AND (IY+d)	CB6D	BIT 5, L	DDE3	EX (SP), IX
		A7	AND A	CB76	BIT 6, (HL)	FDE3	EX (SP), IY
		A0	AND B	DDCB0576	BIT 6, (IX+d)	08	EX AF, AF
		A1	AND C	FDCB0576	BIT 6, (IY+d)	EB	EX DE, HL
		A2	AND D	CB77	BIT 6, A	D9	EXX
		A3	AND E	CB70	BIT 6, B	76	HALT
		A4	AND H	CB71	BIT 6, C	ED46	IM 0
		A5	AND L	CB72	BIT 6, D	ED56	IM 1
		E620	AND N	CB73	BIT 6, E	ED5E	IM 2
		CB46	BIT 0, (HL)	CB74	BIT 6, H	ED78	IN A, (C)
		DDCB0546	BIT 0, (IX + d)	CB75	BIT 6, L	DB20	IN A, (N)
		FDCB0546	BIT 0, (IY + d)	CB7E	BIT 7, (HL)	ED40	IN B, (C)
		CB47	BIT 0, A	DDCB057E	BIT 7, (IX+d)	ED48	IN C, (C)
		CB40	BIT 0, B	FDCB057E	BIT 7, (IY+d)	ED50	IN D, (C)
		CB41	BIT 0, C	CB7F	BIT 7, A	ED58	IN E, (C)
		CB42	BIT 0, D	CB78	BIT 7, B	ED60	IN H, (C)
		CB43	BIT 0, E	CB79	BIT 7, C	ED68	IN L, (C)
		CB44	BIT 0, H	CB7A	BIT 7, D	34	INC (HL)
		CB45	BIT 0, L	CB7B	BIT 7, E	DD3405	INC (IX+d)
		CB4E	BIT 1, (HL)	CB7C	BIT 7, H	FD3405	INC (IY+d)
		DDCB054E	BIT 1, (IX+d)	CB7D	BIT 7, L	3C	INC A
		FDC8054E	BIT 1, (IY+d)	DC8405	CALL C, NN	04	INC B
		CB4F	BIT 1, A	FC8405	CALL M, NN	03	INC BC
		CB48	BIT 1, B	D48405	CALL NC, NN	0C	INC C
		CB49	BIT 1, C	CD8405	CALL NN	14	INC D
		CB4A	BIT 1, D	C48405	CALL NZ, NN	13	INC DE
		CB4B	BIT 1, E	F48405	CALL P, NN	1C	INC E
		CB4C	BIT 1, H	EC8405	CALL PE, NN	24	INC H
		CB4D	BIT 1, L	E48405	CALL PO, NN	23	INC HL
		CB56	BIT 2, (HL)	CC8405	CALL Z, NN	DD23	INC IX
		DDC80556	BIT 2, (IX+d)	3F	CCF	FD23	INC IY
		FDCB0556	BIT 2, (IY+d)	BE	CP (HL)	2C	INC L
		CB57	BIT 2, A	DDBE05	CP(IX+d)	33	INC SP
		CB50	BIT 2, B	FDBE05	CP (IY+d)	EDAA	IND
		CB51	BIT 2, C	BF	CP A	EDBA	INDR
		CB52	BIT 2, D	B8	CP B	EDA2	INI
		CB53	BIT 2, E	B9	CP C	EDB2	INIR
		CB54	BIT 2, H	BA	CP D	E9	JP (HL)
		CB55	BIT 2, L	BB	CP E	DDE9	JP (IX)
		CB5E	BIT 3, (HL)	BC	CP H	FDE9	JP (IY)
		DDCB055E	BIT 3, (IX+d)	BD	CP L	DA8405	JP C, NN
		FDCB055E	BIT 3, (IY+d)	FE20	CP N	FA8405	JP M, NN
		CB5F	BIT 3, A	EDA9	CPD	D28405	JP NC, NN
		CB58	BIT 3, B	EDB9	CPDR	C38405	JP NN

Z80-CPU register configuration

Special purpose registers

Interrupt Vector I	Memory Refresh R
Index register IX	
Index register IY	
Stack pointer SP	
Program counter PC	

General purpose registers

Main reg set:
Accumulator A	Flags F	B	C	D	E	H	L

Alternate reg set:
Accumulator A	Flags F	B	C	D	E	H	L

Obj code	Source statement	Obj code	Source statement	Obj code	Source Statement	Obj code	Source statement
C28405	JP, NZ, NN	45	LD B, L	ED4F	LD R, A	DDCB0596	RES 2, (IX+d)
F28405	JP P, NN	0620	LD B, N	ED7B8405	LD SP, (NN)	FDCB0596	RES 2, (IY+d)
EA8405	JP PE, NN	ED4B8405	LD BC, (NN)	F9	LD SP, HL	CB97	RES 2, A
E28405	JP PO, NN	018405	LD BC, NN	DDF9	LD SP, IX	CB90	RES 2, B
CA8405	JP Z, NN	4E	LD C, (HL)	FDF9	LD SP, IY	CB91	RES 2, C
382E	JR C, DIS	DD4E05	LD C, (IX+d)	318405	LD SP, NN	CB92	RES 2, D
182E	JR DIS	FD4E05	LD C, (IY+d)	EDA8	LDD	CB93	RES 2, E
302E	JR NC, DIS	4F	LD C, A	EDB8	LDDR	CB94	RES 2, H
202E	JR NZ, DIS	48	LD C, B	EDA0	LDI	CB95	RES 2, L
282E	JR Z, DIS	49	LD C, C	EDB0	LDIR	CB9E	RES 3, (HL)
02	LD (BC), A	4A	LD C, D	ED44	NEG	DDCB059E	RES 3, (IX+d)
12	LD (DE), A	4B	LD C, E	00	NOP	FDCB059E	RES 3, (IY+d)
77	LD (HL), A	4C	LD C, H	B6	OR (HL)	CB9F	RES 3, A
70	LD (HL), B	4D	LD C, L	DDB605	OR (IX+d)	CB98	RES 3, B
71	LD (HL), C	0E20	LD C, N	FDB605	OR (IY+d)	CB99	RES 3, C
72	LD (HL), D	56	LD D, (HL)	B7	OR A	CB9A	RES 3, D
73	LD (HL), E	DD5605	LD D, (IX+d)	B0	OR B	CB9B	RES 3, E
74	LD (HL), H	FD5605	LD D, (IY+d)	B1	OR C	CB9C	RES 3, H
75	LD (HL), L	57	LD D, A	B2	OR D	CB9D	RES 3, L
3620	LD (HL), N	50	LD D, B	B3	OR E	CBA6	RES 4, (HL)
DD7705	LD (IX+d), A	51	LD D, C	B4	OR H	DDCB05A6	RES 4, (IX+d)
DD7005	LD (IX+d), B	52	LD D, D	B5	OR L	FDCB05A6	RES 4, (IY+d)
DD7105	LD (IX+d), C	53	LD D, E	F620	OR N	CBA7	RES 4, A
DD7205	LD (IX+d), D	54	LD D, H	EDB	OTDR	CBA0	RES 4, B
DD7305	LD (IX+d), E	55	LD D, L	ED83	OTIR	CBA1	RES 4, C
DD7405	LD (IX+d), H	1620	LD D, N	ED79	OUT (C), A	CBA2	RES 4, D
DD7505	LD (IX+d), L	ED5B8405	LD DE (NN)	ED41	OUT (C), B	CBA3	RES 4, E
DD360520	LD (IX+d), N	118405	LD DE, NN	ED49	OUT (C), C	CBA4	RES 4, H
FD7705	LD (IY+d), A	5E	LD E, (HL)	ED51	OUT (C), D	CBA5	RES 4, L
FD7005	LD (IY+d), B	DD5E05	LD E, (IX+d)	ED59	OUT (C), E	CBAE	RES 5, (HL)
FD7105	LD (IY+d), C	FD5E05	LD E, (IY+d)	ED61	OUT (C), H	DDCB05AE	RES 5, (IX+d)
FD7205	LD (IY+d), D	5F	LD E, A	ED69	OUT (C), L	FDCB05AE	RES 5, (IY+d)
FD7305	LD (IY+d), E	58	LD E, B	D320	OUT (N), A	CBAF	RES 5, A
FD7405	LD (IY+d), H	59	LD E, C	EDAB	OUTD	CBA8	RES 5, B
FD7505	LD (IY+d), L	5A	LD E, D	EDA3	OUTI	CBA9	RES 5, C
FD360520	LD (IY+d), N	5B	LD E, E	F1	POP AF	CBAA	RES 5, D
328405	LD (NN), A	5C	LD E, H	C1	POP BC	CBAB	RES 5, E
ED438405	LD (NN), BC	5D	LD E, L	D1	POP DE	CBAC	RES 5, H
ED538405	LD (NN), DE	1E20	LD E, N	E1	POP HL	CBAD	RES 5, L
228405	LD (NN), HL	66	LD H, (HL)	DDE1	POP IX	CBB6	RES 6, (HL)
DD228405	LD (NN), IX	DD6605	LD H, (IX+d)	FDE1	POP IY	DDCB05B6	RES 6, (IX+d)
FD228405	LD (NN), IY	FD6606	LD H, (IY+d)	F5	PUSH AF	FDCB05B6	RES 6, (IY+d)
ED738405	LD (NN), SP	67	LD H, A	C5	PUSH BC	CBB7	RES 6, A
0A	LD A, (BC)	60	LD H, B	D5	PUSH DE	CBB0	RES 6, B
1A	LD A, (DE)	61	LD H, C	E5	PUSH HL	CBB1	RES 6, C
7E	LD A, (HL)	62	LD H, D	DDE5	PUSH IX	CBB2	RES 6, D
DD7E05	LD A, (IX+d)	63	LD H, E	FDE5	PUSH IY	CBB3	RES 6, E
FD7E05	LD A, (IY+d)	64	LD H, H	CB86	RES 0, (HL)	CBB4	RES 6, H
3A8405	LD A, (NN)	65	LD H, L	DDCB0586	RES 0, (IX+d)	CBB5	RES 6, L
7F	LD A, A	2620	LD H, N	FDCB0586	RES 0, (IY+d)	CBBE	RES 7, (HL)
78	LD A, B	2A8405	LD HL, (NN)	CB87	RES 0, A	DDCB05BE	RES 7, (IX+d)
79	LD A, C	218405	LD HL, NN	CB80	RES 0, B	FDCB05BE	RES 7, (IY+d)
7A	LD A, D	ED47	LD I, A	CB81	RES 0, C	CBBF	RES 7, A
7B	LD A, E	DD2A8405	LD IX, (NN)	CB82	RES 0, D	CBB8	RES 7, B
7C	LD A, H	DD218405	LD IX, NN	CB83	RES 0, E	CBB9	RES 7, C
ED57	LD A, I	FD2A8405	LD IY, (NN)	CB84	RES 0, H	CBBA	RES 7, D
7D	LD A, L	FD218405	LD IY, NN	CB85	RES 0, L	CBBB	RES 7, E
3E20	LD A, N	6E	LD L, (HL)	CB8E	RES 1, (HL)	CBBC	RES 7, H
ED5F	LD A, R	DD6E05	LD L, (IX+d)	DDCB058E	RES 1, (IX+d)	CBBD	RES 7, L
46	LD B, (HL)	FD6E05	LD L, (IY+d)	FDCB058E	RES 1, (IY+d)	C9	RET
DD4605	LD B, (IX+d)	6F	LD L, A	CB8F	RES 1, A	D8	RET C
FD4605	LD B, (IY+d)	68	LD L, B	CB88	RES 1, B	F8	RET M
47	LD B, A	69	LD L, C	CB89	RES 1, C	D0	RET NC
40	LD B, B	6A	LD L, D	CB8A	RES 1, D	C0	RET NZ
41	LD B, C	6B	LD L, E	CB8B	RES 1, E	F0	RET P
42	LD B, D	6C	LD L, H	CB8C	RES 1, H	E8	RET PE
43	LD B, E	6D	LD L, L	CB8D	RES 1, L	E0	RET PO
44	LD B, H	2E20	LD L, N	CB96	RES 2, (HL)	C8	RET Z

Obj code	Source statement	Obj code	Source statement	Obj code	Source Statement
ED4D	RETI	ED52	SBC HL, DE	CBF0	SET 6, B
ED45	RETN	ED62	SBC HL, HL	CBF1	SET 6, C
CB16	RL (HL)	ED72	SBC HL, SP	CBF2	SET 6, D
DDCB0516	RL (IX+d)	37	SCF	CBF3	SET 6, E
FDCB0516	Rl (IY+d)	CBC6	SET 0, (HL)	CBF4	SET 6, H
CB17	RL A	DDCB05C6	SET 0, (IX+d)	CBF5	SET 6, (HL)
CB10	RL B	FDCB05C6	SET 0, (IY+d)	CBFE	SET 6, L
CB11	RL C	CBC7	SET 0, A	DDCB05FE	SET 7, (IX+d)
CB12	RL D	CBC0	SET 0, B	FDCB05FE	SET 7, (IY+d)
CB13	RL E	CBC1	SET 0, C	CBFF	SET 7, A
CB14	RL H	CBC2	SET 0, D	CBF8	SET 7, B
CB15	RL L	CBC3	SET 0, E	CBF9	SET 7, C
17	RL A	CBC4	SET 0, H	CBFA	SET 7, D
CB06	RLC (HL)	CBC5	SET 0, L	CBFB	SET 7, E
DDCB0506	RLC (IX+d)	CBCE	SET 1, (HL)	CBFC	SET 7, H
FDCB0506	RLC (IY+d)	DDCB05CE	SET 1, (IX+d)	CBFD	SET 7, L
CB07	RLC A	FDCB05CE	SET 1, (IY+d)	CB26	SLA (HL)
CB00	RLC B	CBCF	SET 1, A	DDCB0526	SLA (IX+d)
CB01	RLC C	CBC8	SET 1, B	FDCB0526	SLA (IY+d)
CB02	RLC D	CBC9	SET 1, C	CB27	SLA A
CB03	RLC E	CBCA	SET 1, D	CB20	SLA B
CB04	RLC H	CBCB	SET 1, E	CB21	SLA C
CB05	RLC L	CBCC	SET 1, H	CB22	SLA D
07	RLCA	CBCD	SET 1, L	CB23	SLA E
ED6F	LRD	CBD6	SET 2, (HL)	CB24	SLA H
CB1E	RR (HL)	DDCB05D6	SET 2, (IX+d)	CB25	SLA L
DDCB051E	RR (IX+d)	FDCB05D6	SET 2, (IY+d)	CB2E	SRA (HL)
FDCB051E	RR (IY+d)	CBD7	SET 2, A	DDCB052E	SRA (IX+d)
CB1F	RR A	CBD0	SET 2, B	FDCB052E	SRA (IX+d)
CB18	RR B	CBD1	SET 2, C	CB2F	SRA A
CB19	RR C	CBD2	SET 2, D	CB28	SRA B
CB1A	RR D	CBD3	SET 2, E	CB29	SRA C
CB1B	RR E	CBD4	SET 2, H	CB2A	SRA D
CB1C	RR H	CBD5	SET 2, L	CB2B	SRA E
CB1D	RR L	CBDE	SET 3, (HL)	CB2C	SRA H
IF	RRA	DDCB05DE	SET 3, (IX+d)	CB2D	SRA L
CB0E	RRC (HL)	FDCB05DE	SET 3, (IY+d)	CB3E	SRL (HL)
DDCB050E	RRC (IX+d)	CBDF	SET 3, A	DDCB053E	SRL (IX+d)
FDCB050E	RRC (IY+d)	CBD8	SET 3, B	FDCB053E	SRL (IY+d)
CB0F	RRC A	CBD9	SET 3, C	CB83F	SRL A
CB08	RRC B	CBDA	SET 3, D	CB38	SRL B
CB09	RRC C	CBDB	SET 3, E	CB39	SRL C
CB0A	RRC D	CBDC	SET 3, H	CB3A	SRL D
CB0B	RRC E	CBDD	SET 3, L	CB3B	SRL E
CB0C	RRC H	CBE6	SET 4, (HL)	CB3C	SRL H
CB0D	RRC L	DDCB05E6	SET 4, (IX+d)	CB3D	SRL L
0F	RRCA	FDCB05E6	SET 4, (IY+d)	96	SUB (HL)
ED67	RRD	CBE7	SET 4, A	DD9605	SUB (IX+d)
C7	RST 0	CBE0	SET 4, B	FD9605	SUB (IY+d)
D7	RST 10H	CBE1	SET 4, C	97	SUB A
DF	RST 18H	CBE2	SET 4, D	90	SUB B
E7	RST 20H	CBE3	SET 4, E	91	SUB C
EF	RST 28H	CBE4	SET 4, H	92	SUB D
F7	RST 30H	CBE5	SET 4, L	93	SUB E
FF	RST 38H	CBEE	SET 5, (HL)	94	SUB H
CF	RST 8	DDCB05EE	SET 5, (IX+d)	95	SUB L
9E	SBC A, (HL)	FDCB05EE	SET 5, (IY+d)	D620	SUB N
DD9E05	SBC A, (IX+d)	CBEF	SET 5, A	AE	XOR (HL)
FD9E05	SBC A, (IY+d)	CBE8	SET 5, B	DDAE05	XOR (IX+d)
9F	SBC A, A	CBE9	SET 5, C	FDAE05	XOR (IY+d)
98	SBC A, B	CBEA	SET 5, D	AF	XOR A
99	SBC A, C	CBEB	SET 5, E	A8	XOR B
9A	SBC A, D	CBEC	SET 5, H	A9	XOR C
9B	SBC A, E	CBED	SET 5, L	AA	XOR D
9C	SBC A, H	CBF6	SET 6, (HL)	AB	XOR E
9D	SBC A, L	DDCB05F6	SET 6, (IX+d)	AC	XOR H
DE20	SBC A, N	FDCB05F6	SET 6, (IY+d)	AD	XOR L
ED42	SBC HL, BC	CBF7	SET 6, A	EE20	XOR N

Appendix D

PIO: Parallel Input/Output

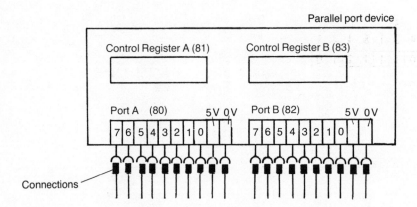

NOTES: (a) CR is control register
         (b) Figures in brackets are typical addresses
         (c) All numbers are in hexadecimal

Mode	Description	Contents of Control Register
0	Whole port as outputs	0F
1	Whole port as inputs	4F
2	Bidirectional	8F Port A only
3	Mixed inputs/outputs	First CF, followed by the 8-bit word, 1 for input 0 for output

EXAMPLES

	Assembly language	Mode
1. Port A all inputs	LDA,4F OUT (81),A	1
2. Port B mixed	LDA,CF OUT (83),A LDA,3E OUT (83),A	3

```
        3       E
  8  4  2  1  8  4  2  1
┌──┬──┬──┬──┬──┬──┬──┬──┐
│0 │0 │1 │1 │1 │1 │1 │0 │
└──┴──┴──┴──┴──┴──┴──┴──┘
```

Appendix E

Bit Handling of PIO Ports using Machine Code

One way is via the accumulator. The table gives the coding.

Assembly instr.	Accumulator bit no.							
	7	6	5	4	3	2	1	0
SET N,A	CBFF	CBF7	CBEF	CBE7	CBDF	CBD7	CBCF	CBC7
RES N,A	CBBF	CBB7	CBAF	CBA7	CB9F	CB97	CB8F	CB87
BIT N,A	CB7F	CB77	CB6F	CB67	CB5F	CB57	CB4F	CB47

To set or reset bits of a port:

1. copy the port contents into the accumulator;
2. set or reset the required bits;
3. copy the accumulator contents back to the port.

EXAMPLE: Reset bit 4 on port A

DB	IN A,(80)
80	
CB	RES 4,A
A7	
D3	OUT (80),A
80	

To read bits of a port:

1. copy the port contents into the accumulator:
2. read the bits required.

NOTE: 0 is taken as zero in a logic instruction.

EXAMPLE: Read bit 6 on port B

DB	IN A,(81)
81	
CB	BIT 6,A
77	

Appendix F

Bit Handling of PIO Ports Using BASIC

To set and reset bits using BASIC it is necessary to use the logical AND and OR statements.

1 AND 1 = 1	1 OR 1 = 0
1 AND 0 = 0	1 OR 0 = 1
0 AND 1 = 0	0 OR 1 = 1
0 AND 0 = 0	0 OR 0 = 0

The usual requirement is to set and reset individual bits, without affecting the remainder, or to be able to read individual bits.

1. To Set a Bit of a Port
OR the existing port configuration with a mask of zeros except from the bit to be set which will be a 1.

EXAMPLE

1	1	0	×	0	1	1	0

bit × to be set

0 0 0 1 0 0 0 0   OR the mask

1	1	0	1	0	1	1	0

result, only the required bit is changed

2. To Reset a Bit of a Port
AND the existing port configuration with a mask of all ones except from the bit to be reset which will be a zero.

EXAMPLE

1	1	0	0	1	0	×	0

bit × to be set

1 1 1 1 1 1 0 1   AND the mask

1	1	0	0	1	0	0	0

result, only the required bit is changed

3. To Read a Bit of a Port

AND the existing port configuration with a mask of all ones except from the bit to be reset which will be a zero.

EXAMPLE

1	×	1	0	0	1	1	1

bit × to be set

0 1 0 0 0 0 0 0   AND the mask

0	×	0	0	0	0	0	0

result, if the port bit is a one then the result is read as 1; if the bit is zero it is read as 0.

Appendix G

Pulse Width Modulation

Use for the Control of Solenoids

It is possible to raise the force produced by a solenoid by increasing the input voltage to it, but to switch on, say, double the voltage rating of a solenoid, and leave it switched on, would cause the solenoid quickly to burn out. This can be overcome by using pulse width modulation (PWM) switching techniques.

The normal switching frequency for pulse width modulation is between 2 and 10 kHz.

For example, by rapidly switching a 24 V supply on and off, for equal time periods, an average output voltage of 12 V is produced.

The time that the voltage is switched on is known as the 'mark' and the time that the voltage is switched off is called the 'space'. If the ratio between the mark and the space (known as the mark–space ratio) is varied, the average voltage alters accordingly. This principle is shown in Fig. G.1

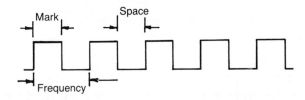

Fig. G.1   Pulses of equal duration on and off.

By maintaining a constant peak voltage of 24 V, and varying the mark–space ratio between 1 : 100 and 1 : 1, the average voltage output would vary between 1% and 100% of the 12 V control voltage.

For example, if a solenoid was acting against a valve spool which in turn was acting against a spring, the spool would move against the spring in direct proportion to the input force provided by the solenoid (Hooke's Law). This holds true for springs with more than three full turns. Some spool valves have springs with less than three turns. As fluid flows through the valve in direction relationship to the spool movement, any variation in the solenoid force due to changes in the average voltage will give a different output flow.

There are two main ways of achieving pulse width modulation:

1.   a fixed mark with a varying space;
2.   fixing the space and varying the mark.

In the first case, varying the mark here gives pulse width modulation. However, here the cycle time or frequency remains constant.

In the second case, fixing the mark or time that the voltage is switched on and varying the cycle time or frequency achieves PWM.

These two methods are illustrated in Figs. G.2 and G.3 with 1 : 1 and 1 : 3 mark–space ratios. The two examples are superimposed.

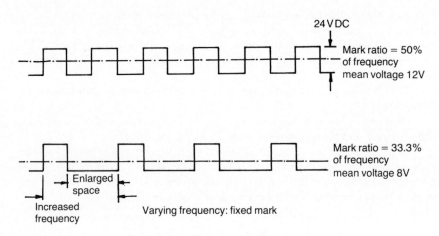

Fig. G.2   Fixed mark: varying space.

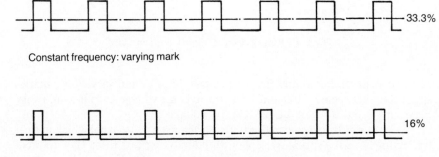

Fig. G.3   Constant frequency: varying mark.

Using a microprocessor, it is quite simple to generate pulse trains to provide a controlled voltage.

Microcomputer Control of Pulse Width Modulation

The microprocessor can produce a pulse modulated signal in two ways, software control or through support chips. Software control requires constant attention by the microprocessor and therefore means that the microprocessor cannot be used for other functions.

Using hardware support chips will relieve this monopolization of the microprocessor because it will only update the support chip when changes are necessary. This is the most practical way of operating a microprocessor system and is the one recommended.

A support chip used with the Z80 microprocessor is the counter timer circuit (CTC).

Index